VOLUME 1

Argus Books
Wolsey House
Wolsey Road
Hemel Hempstead
Herts. HP2 4SS
England

First published by Argus Books 1989

ISBN 0 85242 983 5

Designed by Little Oak Studios
Phototypesetting by Typesetters (Birmingham) Ltd
Printed and bound in Great Britain by
William Clowes Limited, Beccles and London

Cover photo: Nieuport 28C-1 restoration in the Cole Palen collection at Old Rhinebeck is operated by Doug Campbell, has acetylene gas guns and has appeared (as here) at hard runway airports as well as its narrow strip through the woods home base.

AIRCRAFT ARCHIVE

AIRCRAFT OF WORLD WAR ONE

VOLUME 1

Contents

A DETAILED COLLECTION OF ORIGINAL SCALE AIRCRAFT DRAWINGS

Introduction

Interest in the comparatively fragile aeroplanes which fought over European skies in the Great War of 1914–1918 has, like tides over the broad sandbanks of the Somme estuary, drifted back and forth over the past 75 years to varying depths of intensity.

Through films, the specialist magazines and latter-day TV documentaries, the pioneering spirit of those early aerial gladiators has been robustly sustained. Museums, builders of replicas and models have advanced the levels of research to incredible heights of authenticity. Even today, those rabid enthusiasts to whom the period is an episode of modern history, but outside their own lifetime, are still uncovering hidden documentation on the men and machines involved in that 'War to end all wars'.

This burst of fervour for information so long after those years which witnessed the emergence of the aeroplane from a primitive frame to a practical transport has been nurtured by dedicated publications such as *Cross & Cockade, World War One Aero* and *Windsock International.* In a way, they reflect a publishing period of the 1930s, but to a degree of detail that would astound the authors in *Popular Flying, Air Stories* or *Flying Aces.*

We owe so much to W E John's fictional 'Captain James Biggles' and Donald Keyhoe's 'Philip Strange', and to the thrilling personal stories by Arch Whitehouse and Joe Archibald. Illustrators Stanley Orton Bradshaw and Howard Leigh recaptured air battle scenes while still fresh in the memory, and those scale drawing pioneers Leonard Bridgeman and James Hay Stevens made accurate models possible for the first time.

Parallel with the publishers were those rare British individuals Richard Shuttleworth and R G J Nash, who sought to preserve WWI aeroplanes for posterity. It is thanks to them that visitors to the Shuttleworth Collection and the Royal Air Force Museum can gaze upon the actual hardware, restored to perfection and, in the case of Shuttleworth, in flying condition.

Twenty-one peaceful years erupted into World War II, and with it another surge of aero interest, but this time into the topical with a new breed of air hero and his now inevitably monoplane mount.

Memories fade with age, and the events of 1939–1945 tended to eclipse earlier happenings. The airfields of Izel le Hameau, Bruay, Vert Galland Estaires or Hesdigneul would no longer suit the powerful machines which had developed so rapidly. In consequence we

◄ High over the south-east coast of England, a flight of four BE2es is silhouetted in loose formation.

◄ Pfalz D XII postwar, one of those used in Hollywood films, since restored and on show in NASM. This one was restored by Buck Kennel at Paramount in 1938–39 with a false '7517/18' serial.

experienced a dwell in affection for our parachute-less warriors and their gallantry. Until, that is, the late 1950s – which is where this volume opens its collections of drawings.

The following pages carry reprints of 1/72nd scale drawings that appeared in *Aeromodeller* and *Scale Models* over the period from 1958 to 1983. They form a representative collection of work by specialists who introduced new standards of fine-detail draughting. Each has his own style, and all have the same aim – to record with accuracy the shapes and structures of aeroplanes that have in the majority of cases disappeared forever. It is to George Cox, Peter Gray, David Jones, Björn Karlström, Ken McDonough, Felix Pawlowicz, Ian Stair and Harry Woodman that we owe gratitude for the research which enabled them to establish these scale views.

Equally, we are indebted to the photo collections which are credited with each caption. In particular, without the help of Jack Bruce, Alex Imrie, Stuart Leslie, Ray Rimell and Harry Woodman the 100-plus illustrations and their informative descriptions simply would not have been possible. Publishing files are seldom complete after as much as 30 years of storage, and such was the scarcity of original photographs that many had been returned to authors after first appearance in the magazine. Fortunately, friendships which had formed through a mutual enthusiasm for the subject overcame many hurdles and ensured the selection which the reader can now enjoy in this and Volume 2 of the series.

◄ Ground-handling WWI aircraft, without brakes or steerable tailskid, required wing-tip handlers. The Shuttleworth Collection's Sopwith Pup, marked as the Pup prototype, is here being turned on to the Old Warden runway. (R Moulton)

Armstrong Whitworth FK8

Country of origin: Great Britain.
Type: Two-seat corps reconnaissance aircraft.
Powerplant: One Beardmore liquid-cooled engine rated at 120hp or (later version) 160hp.
Dimensions: Wing span 42ft 5in *12.93m*, (160hp engine) 43ft 4in *13.21m*; length 29ft 9in *9.07m*, (160hp engine) 31ft 5in *9.58m*; height 10ft 11in *3.33m*, (160hp engine) 11ft 7in *3.53m*; wing area (160hp engine) 504 sq ft *46.82m²*.
Weights: Empty 1682lb *763kg*, (160hp engine) 1916lb *869kg*; loaded 2447lb *1110kg*, (160hp engine) 2811lb *1275kg*.
Performance: Maximum speed 83.5mph *134.5kph* at 8000ft *2440m*, (160hp engine) 95mph *153kph* at 6500ft *1980m*; time to 10,000ft *3050m*, 35min, (160hp engine) 27.8min; service ceiling 12,000ft *3660m*, (160hp engine) 13,000ft *3960m*; endurance 3hr.
Armament: One fixed Vickers machine gun and one flexibly mounted Lewis machine gun.
Service: First flight May 1916; service entry 1917.

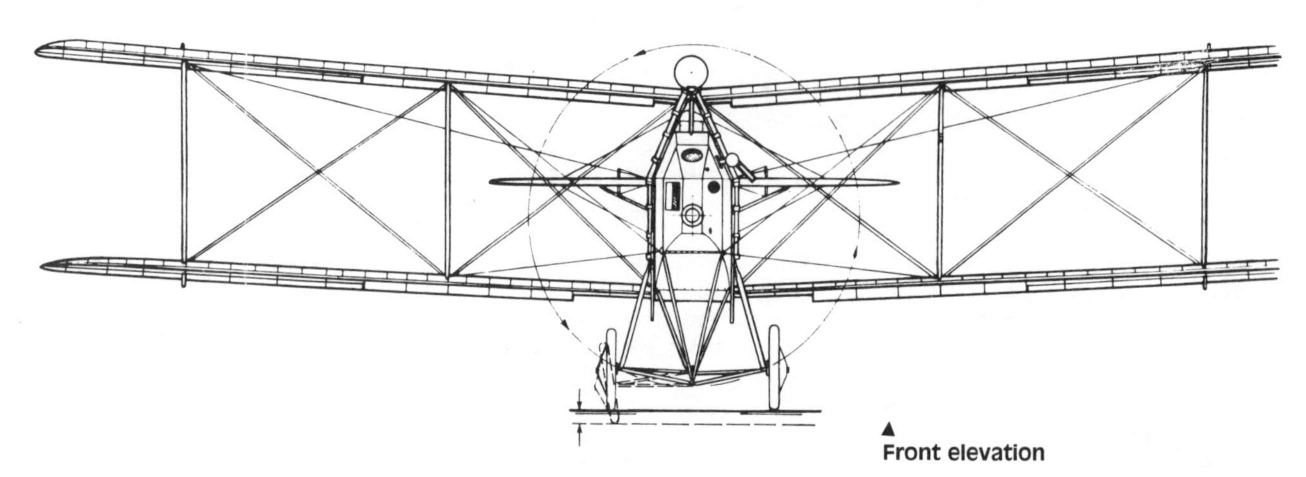

▲
Front elevation

Postwar, FK8s pioneered Australian air routes for the original Quantas Company; regrettably, none survive. (IWM Q69122)
▼

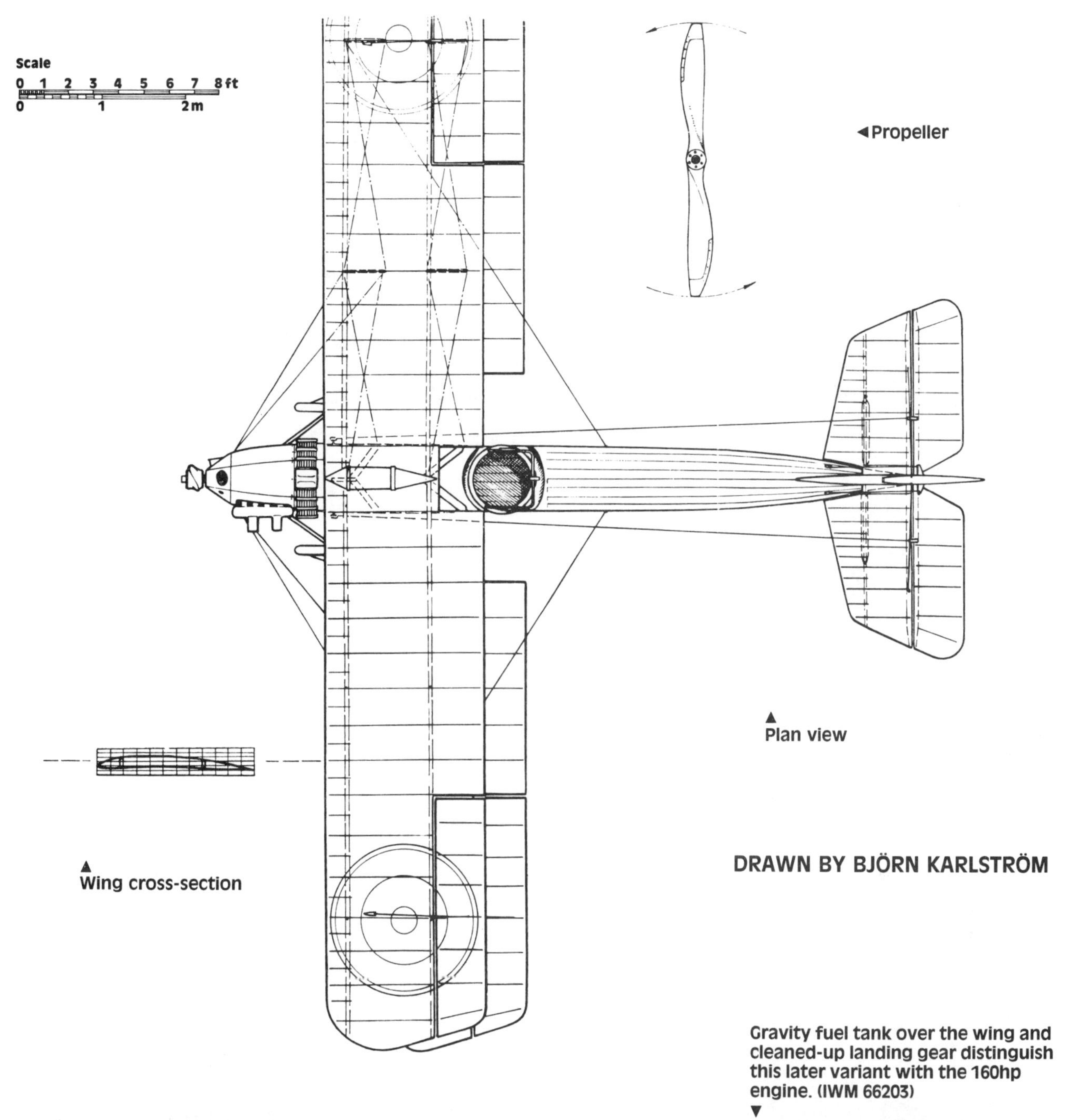

▲
Plan view

▲
Wing cross-section

DRAWN BY BJÖRN KARLSTRÖM

Gravity fuel tank over the wing and cleaned-up landing gear distinguish this later variant with the 160hp engine. (IWM 66203)
▼

▲
B240 has the angular cowling over a 120hp Beardmore engine. Just discernible is the external rod between cockpits for the observer's elevator control. The underside national marking is an unusual red/blue roundel beneath the outer bay. (IWM Q67522)

Scrap starboard elevation
▼

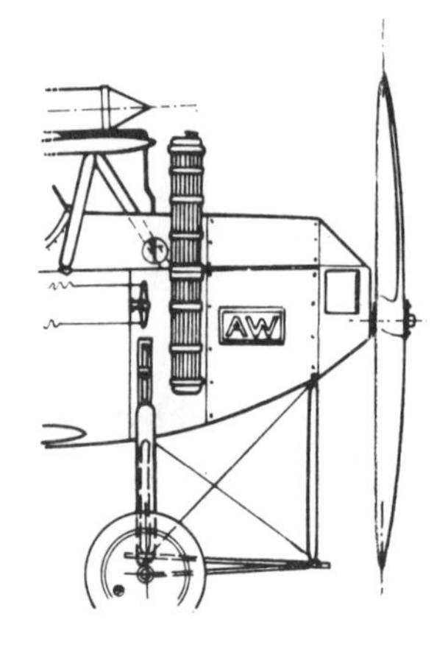

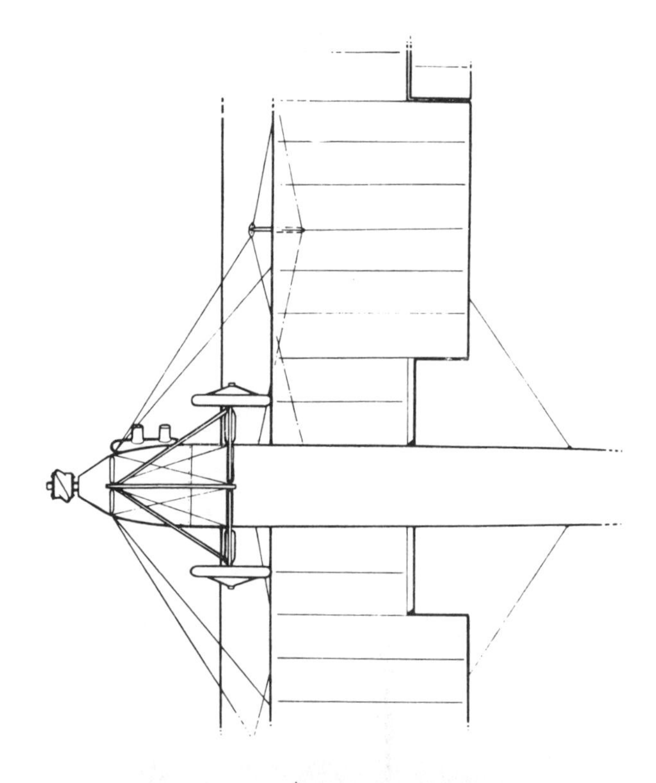

▲
Scrap underplan

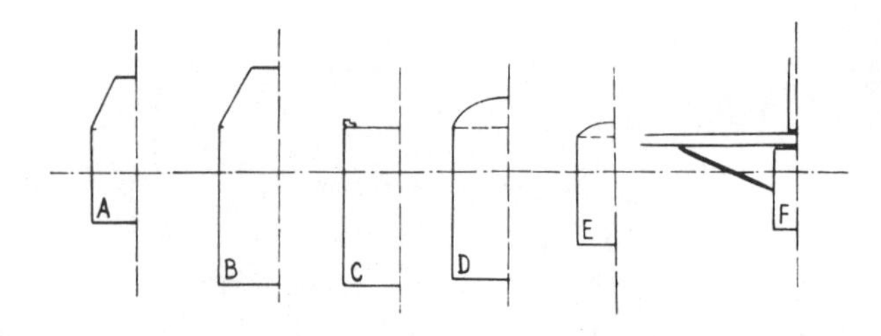

◄ Fuselage cross-sections

Scale

▲
Tip skids on an early FK8, serial A2683. Note atypical centre section for 1916. (IWM 35675)

Port elevation
▼

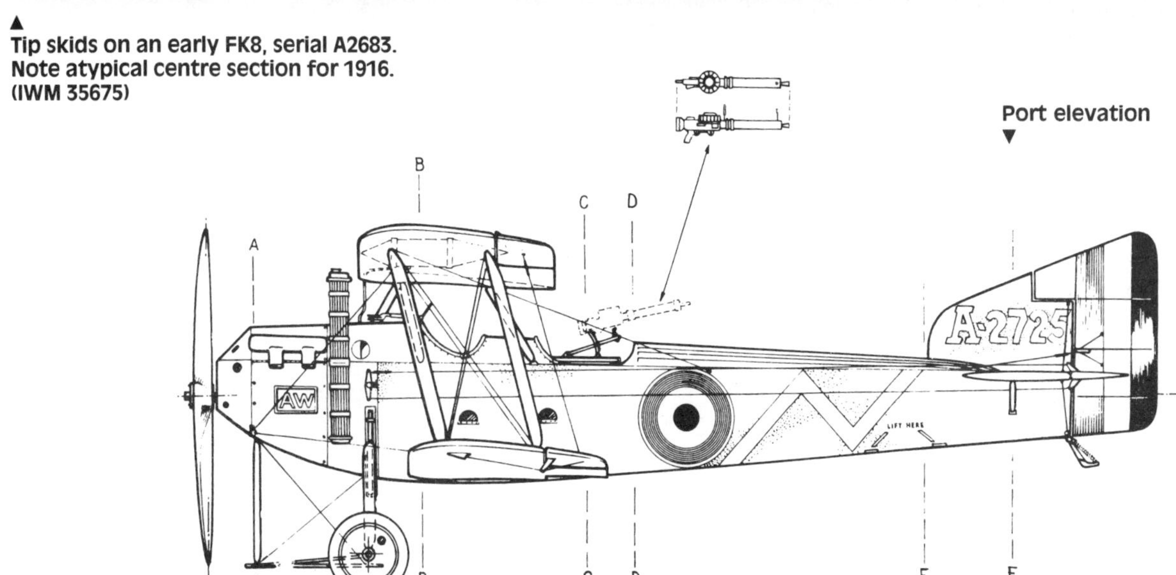

Markings on the FK8 were large as possible! Seven hundred aircraft were in use by the Armistice, its pilots winning two VCs. (IWM Q67068)
▼

Bristol Scout D

Country of origin: Great Britain.
Type: Single-seat scout.
Powerplant: One Gnome or Le Rhône engine rated at 80hp or Monosoupape-Gnome rated at 100hp (also 100hp Clerget and Le Rhône experimental installations).
Dimensions: Wing span 24ft 7in *7.49m*; length 20ft 8in *6.30m*; height 8ft 6in *2.59m*.
Weights: Empty 760lb *345kg*, (100hp engine) 925lb *420kg*; loaded 1250lb *567kg*, (100hp engine) 1440lb *653kg*.
Performance: Maximum speed 100mph *161kph* at ground level, 86mph *138kph* at 10,000ft *3050m*; time to 10,000ft, 18.5min; endurance 2.5hr.
Armament: (Typical) One fixed Vickers or Lewis machine gun.
Service: First flight 1916; service entry 1916.

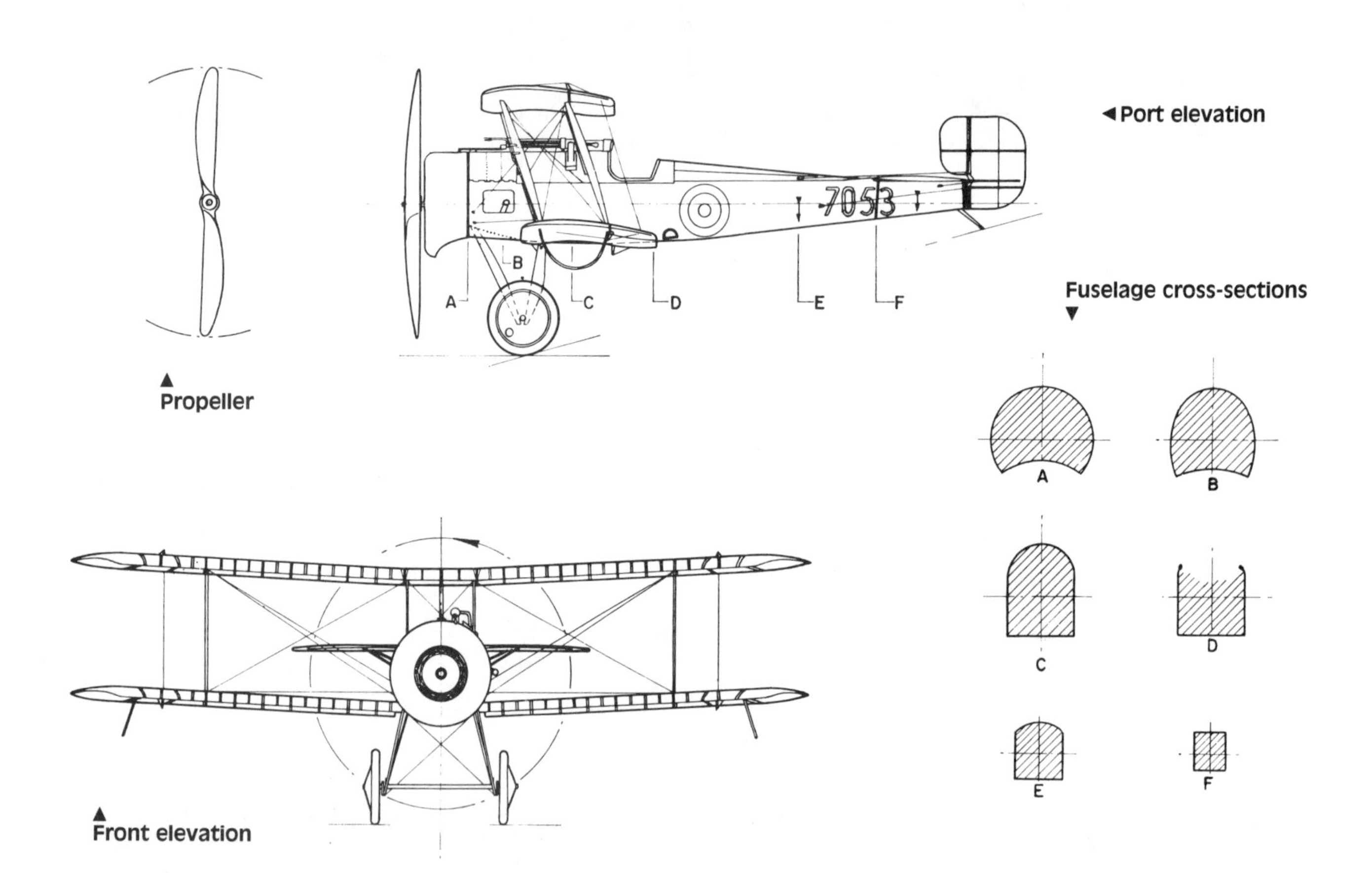

◄ **Lewis MG-armed Scout C stationed at Imbros with the RNAS. The exposed pilot has no windshield or gunsight in this relatively simple, but clean, type. (P L Gray)**

Colour notes
Generally doped, unbleached linen, with natural metal cowling and side panels and varnished wood curved ply decking and struts. Late production aircraft camouflaged on top and side surfaces in khaki-green, with roundels narrowly outlined in white and white serial numbers. Scout Cs of No 2 Wing RNAS at Imbros bore Union Jacks on fuselage instead of roundels, and wing roundels had narrow white outlines on these uncamouflaged aircraft.

DRAWN BY P L GRAY

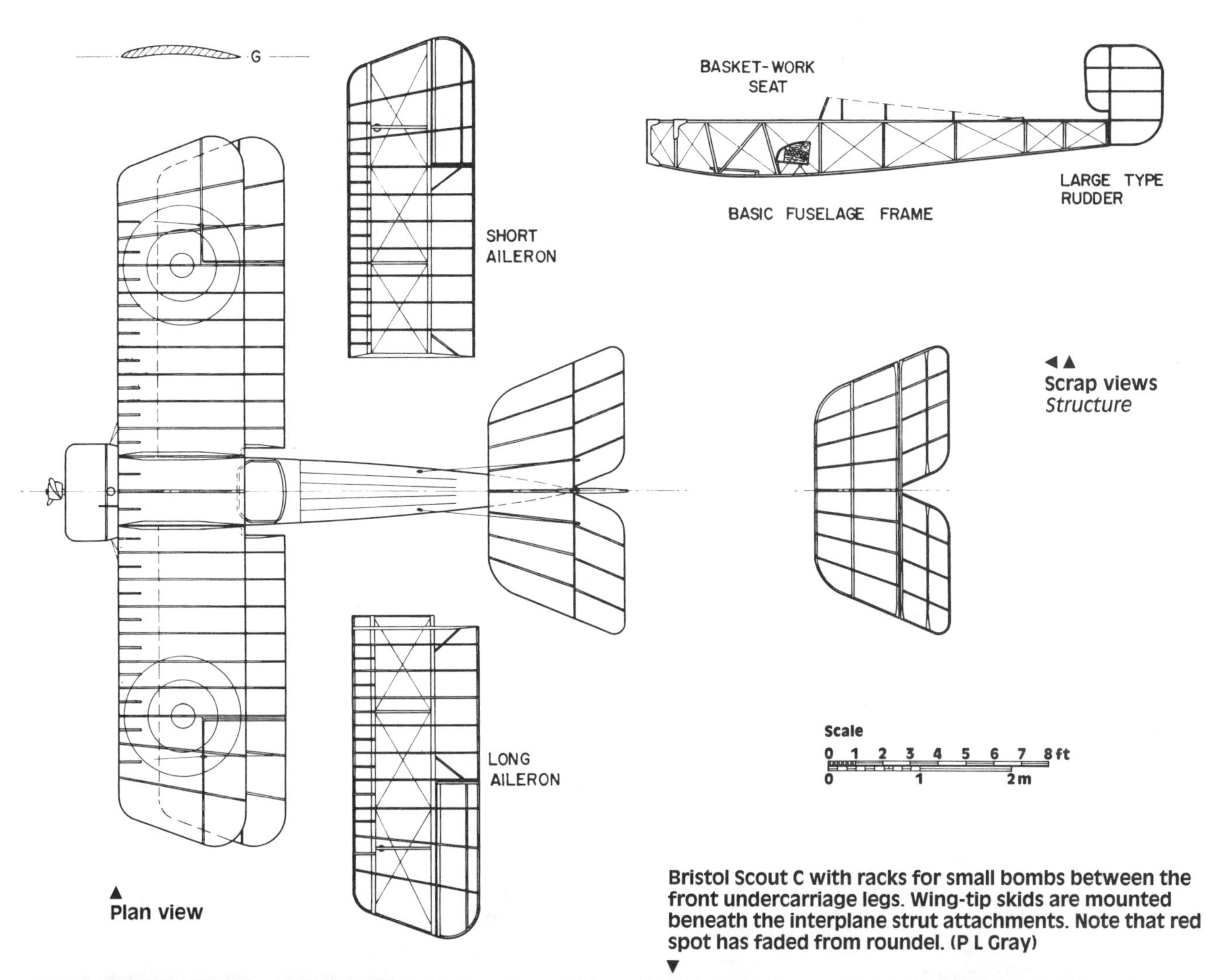

▲
Plan view

◄▲
Scrap views
Structure

Bristol Scout C with racks for small bombs between the front undercarriage legs. Wing-tip skids are mounted beneath the interplane strut attachments. Note that red spot has faded from roundel. (P L Gray)
▼

De Havilland DH4

Country of origin: Great Britain.
Type: Two-seat day bomber.
Powerplant: One RAF 3a engine rated at 200hp, BHP rated at 240hp, Rolls-Royce rated at 250hp or Rolls-Royce Eagle rated at 375hp.
Dimensions: Wing span 42ft 6in *12.95m*; length 29ft 8in *9.04m*, (250hp RR) 30ft 8in *9.35m*, (375hp RR) 30ft 0in *9.14m*; height 10ft 5in *3.17m*, (375hp RR) 11ft 0in *3.35m*; wing area 436 sq ft *40.5m²*.
Weights: Empty 2304lb *1045kg*, (250hp RR) 2303lb *1044kg*, (375hp RR) 2403lb *1090kg*; loaded 3340lb *1515kg*, (250hp RR) 3313lb *1502kg*, (375hp RR) 3472lb *1575kg*.
Performance: Maximum speed 120mph *193kph* at 6500ft *1980m*, (250hp RR) 117mph *188kph* at 6500ft, (375hp RR) 136.5mph *220kph* at 6500ft; time to 6500ft, 8min, (250hp RR) 8.9min, (375hp RR) 5.2min; absolute ceiling 19,500ft *5945m*, (375hp RR) 23,500ft *7165m*; endurance 4hr, (250hp RR) 3.5hr, (375hp RR) 3.75hr.
Armament: Military load 526–545lb *238–247kg*: bombs beneath wings, plus one fixed Vickers machine gun and one single or twin flexibly mounted Lewis gun.
Service: First flight (prototype) August 1916; service entry March 1917.

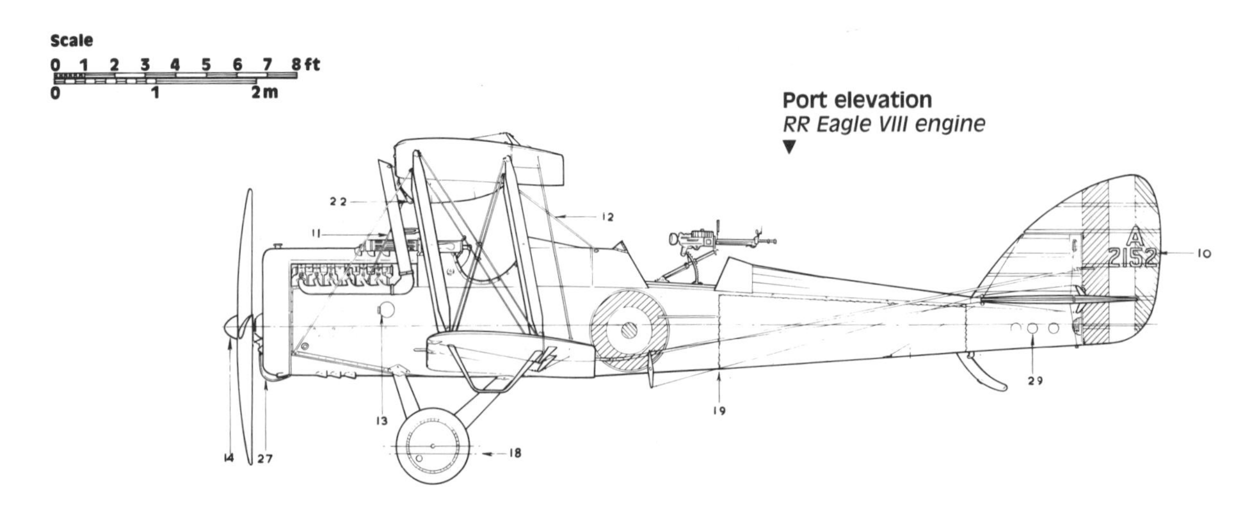

Port elevation
RR Eagle VIII engine

DRAWN BY G A G COX

Windscreen for the rear cockpit is unusual on this DH4. Communication between pilot and observer was limited to a voice tube, and the 60-gallon tank of fuel which separated them led to the 'Flaming Coffin' nickname. (Flight 5398)

◄ **No 27 Squadron RFC paraded 'A' flight at Serny, Pas de Calais, in February 1917. Only the foremost machine, code 'A', has the unpainted cowling. The squadron marking is a vertical white bar. (IWM Q12015)**

Colour notes
Factory finish: Fabric – clear varnish; wood – dark varnish; metal – grey.
Service finish: Khaki-green overall.

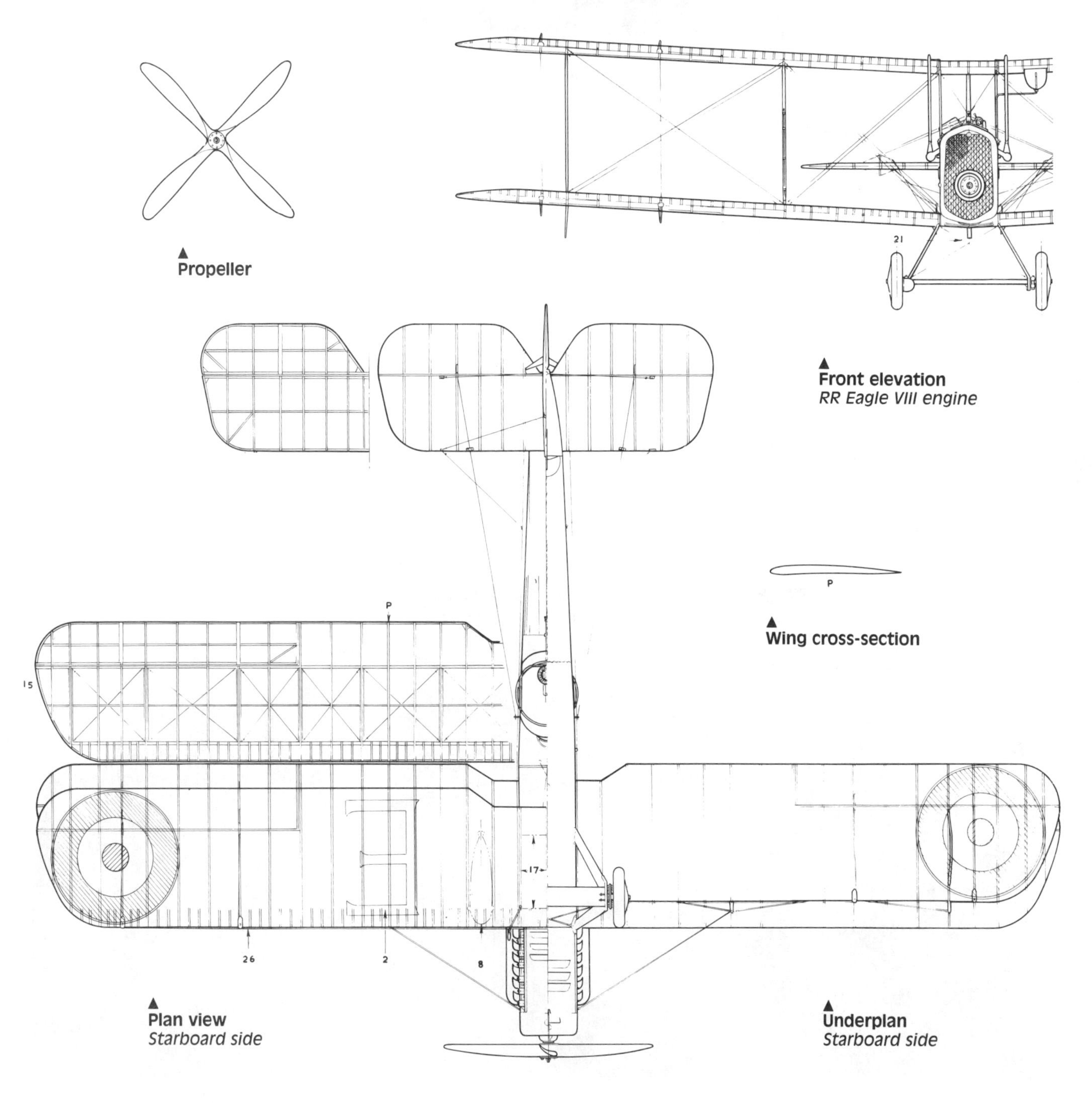

▲ **Propeller**

▲ **Front elevation**
RR Eagle VIII engine

▲ **Wing cross-section**

▲ **Plan view**
Starboard side

▲ **Underplan**
Starboard side

▲
Early-production DH4 A2512 with clear varnished wings and rear fuselage, shorter undercarriage legs and the 250hp Rolls Royce Eagle. Other than rudders stripes, no national markings are carried. (IWM Q56861)

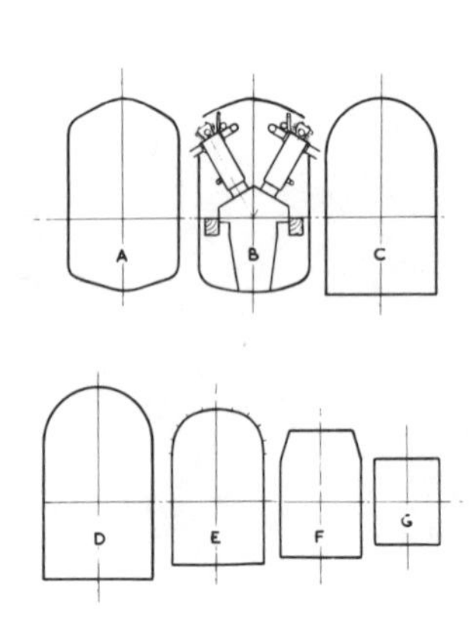

▲
Fuselage cross-sections

Scrap view
Structure
▼

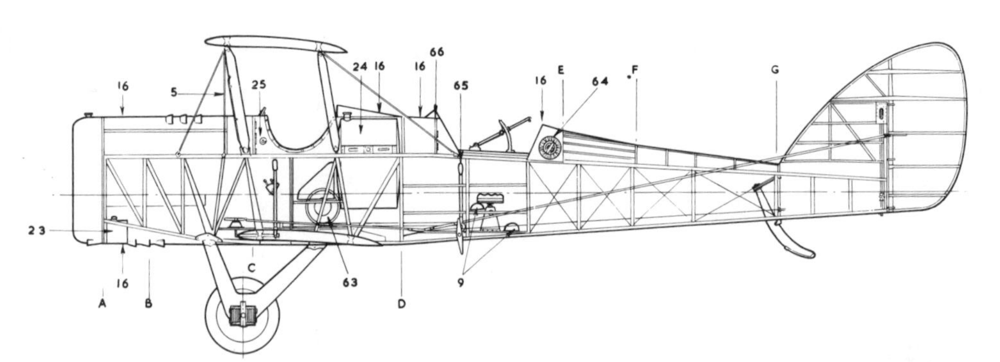

First of the 3106 US DH4s, now with original colours of cream-yellow fuselage, clear doped wings and khaki top decking. (R Moulton)
▼

Following exhibition in the Arts and Industries building and for the US Air Force, the first US-built DH4 awaits restoration at Silver Hill, Washington. (R Moulton)
▼

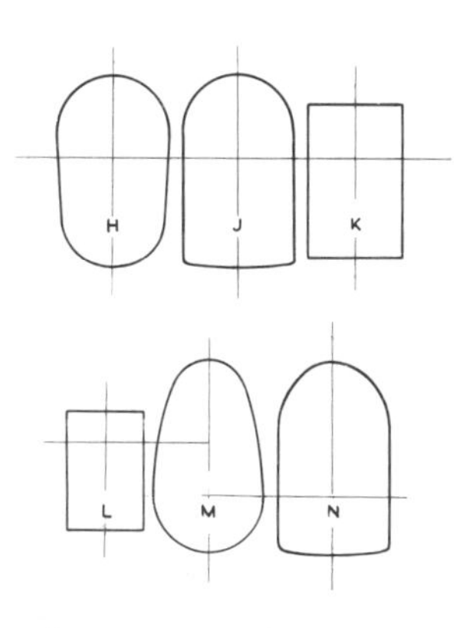

▲
Fuselage cross-sections

▲
No 18 Squadron DH4 with vertical exhaust stack on RAF 3A engine. Underwing bomb racks are loaded. (IWM Q11672)

Scrap front elevation ▼
RAF 3a engine

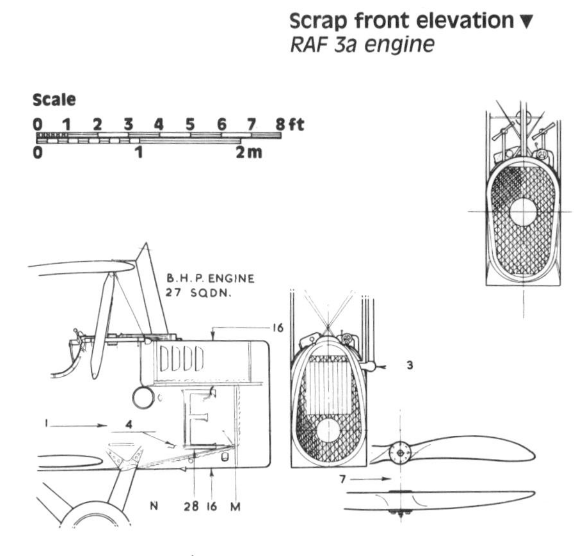

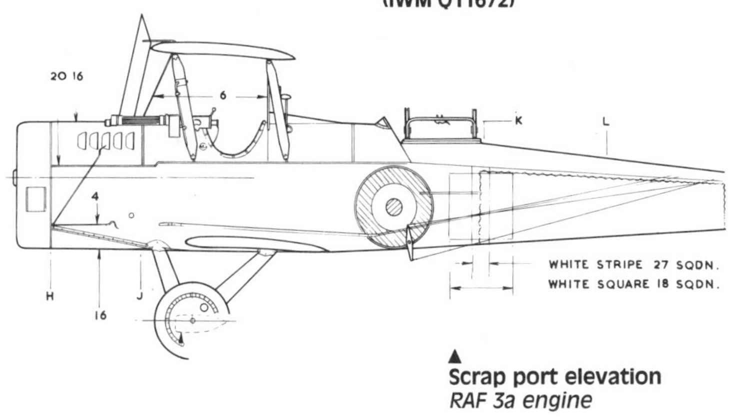

▲
Scrap port elevation
RAF 3a engine

▲
Scrap views
BHP engine

Gunner's cockpit exposed on the Smithsonian DH4, under overhaul at the Paul Garber facility, shows the waist harness and ridged floor panels, plus the infamous between-cockpits fuel tank at right. (R Moulton)
▼

▲
For VIP transport with No 2 (Communication) Squadron, the RAF operated between Kenley and Buc (Paris) with the converted DH4A. Passengers included Winston Churchill and Bonar Law, who requested facing seats for in-flight discussions.

Instone Airlines 'City of Cardiff', converted by A V Roe to a DH4A flew the Croydon–Paris route, was renamed 'City of York' and won the first ever King's Cup Air Race. (E J Riding)
▼

Numerical key

1. BHP-engined versions of 27 Sqn had same rear fuselage as RAF 3a model. **2.** Letter 'E' in white, the same way up on both port and starboard wings. **3.** Extension to exhaust on a few machines only. **4.** Extra wire to lower end of interplane strut. **5.** Extra wire on some Eagle-engined machines. **6.** Centre-section wires crossed on BHP and RAF 3a models. **7.** Two-bladed propeller on BHP engine. **8.** Petrol tank under port wing only. **9.** Steps here on starboard side, all models. **10.** Solid black serial, no white outline. **11.** Parallel wires. **12.** Crossed wires. **13.** Magneto access both sides. **14.** Spinner not always fitted. **15.** Wheel spring covers not always fitted. **16.** Metal panels. **17.** Clear-view panel on some models. **18.** Longer undercarriage on some late Eagle-engined machines. **19.** Plywood covering as far aft as here. **20.** Upper longerons lowered on BHP and RAF 3a models. **21.** Bracing between front legs, but between front and rear on some BHP models. **22.** Petrol gauge. **23.** Oil tank. **24.** Main petrol tank. **25.** Magneto switch. **26.** Aileron wires run along leading edge. **27.** Synchronising gear drive. **28.** White 'E' with black shadow. **29.** Inspection holes

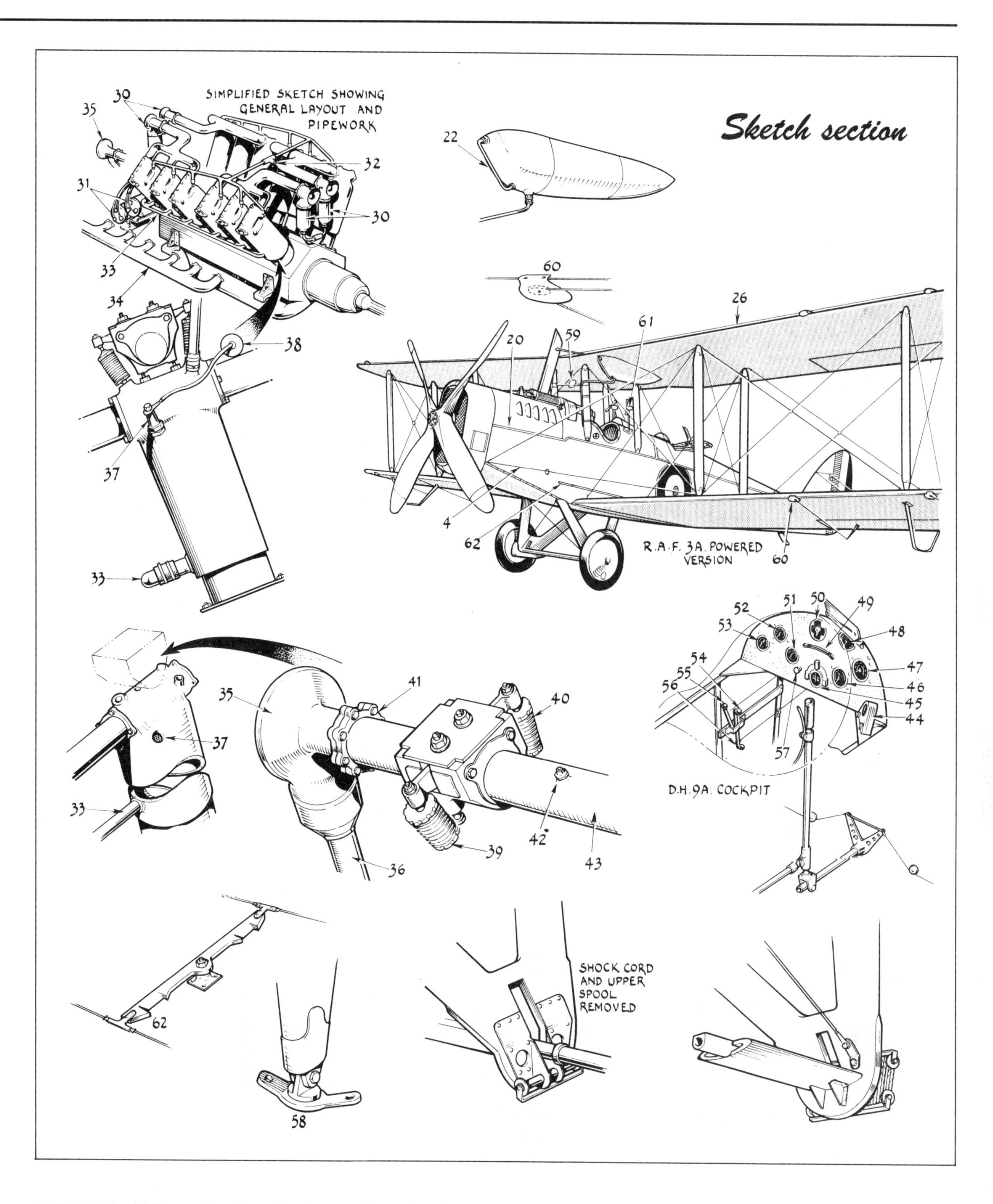

in ply covering. **30.** Two carburettors at each end. **31.** Two magnetos at each side. **32.** Connected to top of radiator. **33.** Connected to lower end of radiator. **34.** Exhausts shown removed. **35.** Bevel gear box for camshaft drive. **36.** Camshaft drive rod housing. **37.** Two plugs to each cylinder, arranged diagonally opposite. **38.** Ignition conduit along top of induction pipes. **39.** Exhaust valve spring. **40.** Induction valve spring. **41.** Camshaft housing coupling between cylinders 2 and 3, 4 and 5. **42.** Grease nipples between cylinders 1 and 2, 3 and 4, 5 and 6. **43.** Camshaft housing. **44.** Delco ignition switches and ammeter. **45.** Illuminated compass. **46.** Air speed indicator. **47.** Altimeter. **48.** Engine rpm. **49.** Lateral level. **50.** Fuel tank selector. **51.** Air pressure. **52.** Radiator temperature. **53.** Oil pressure. **54.** Advance/retard. **55.** Throttle. **56.** Altitude control. **57.** Priming pump. **58.** Typical strut terminal. **59.** Pilot's rear view mirror. **60.** Aileron wire pulley housing. **61.** Petrol pumps. **62.** Rudder bar. **63.** Tail trim wheel. **64.** Ammunition stowage. **65.** Elevator and rudder control for observer. **66.** Observer's instrument panel.

De Havilland DH5

Country of origin: Great Britain.
Type: Single-seat fighter and ground-attack aircraft.
Powerplant: One Le Rhône rotary engine rated at 110hp.
Dimensions: Wing span 25ft 8in *7.82m*; length 22ft 0in *6.71m*; height 9ft 1½in *2.78m*; wing area 212.2 sq ft *19.7m²*.
Weights: Empty 1010lb *458kg*; loaded 1492lb *677kg*.
Performance: Maximum speed 102mph *164kph* at 10,000ft *3050m*; time to 1000ft *305m*, 0.85min; service ceiling 16,000ft *4875m*; endurance 2.75hr.
Armament: One fixed machine gun, plus up to four 25lb *11.3kg* bombs.
Service: First flight 1916; service entry May 1917.

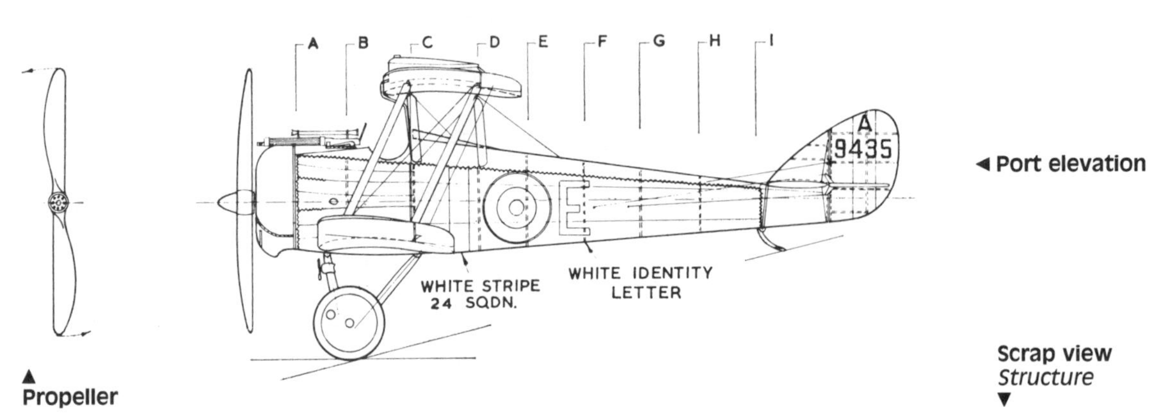

◄ **Port elevation**

▲ **Propeller**

Scrap view
Structure
▼

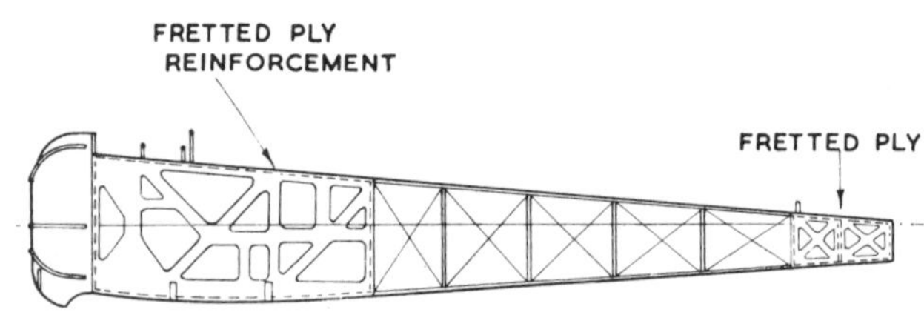

Colour notes
Fabric surfaces were standard khaki-green on top and sides and underneath fuselage; wing and tailplane surfaces were clear doped linen with protective varnish which quickly darkened with age and dirtied with use. Serial numbers painted in black across rudder stripes, usually in fraction fashion if of four digits. Roundels narrowly outlined in white against khaki-green dope.

DRAWN BY P L GRAY

◄ **Bearing the inscription 'Presented by the Native Administration of Benin in South Provinces of Nigeria', A9513 displays its back-stagger and pilot's position ahead of the wing centre section struts. (J M Bruce/G S Leslie)**

◄ **Valued at £2700, A9242 was presented by the women of New South Wales and others in April 1917 and was subsequently known as 'The Women's Battleplane'.**

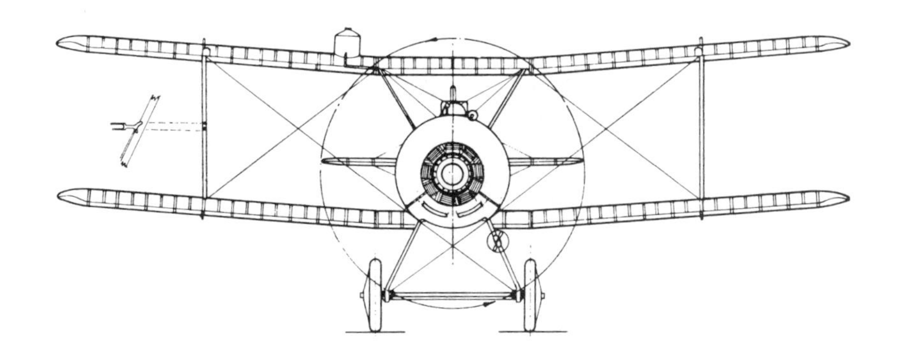

▲ Front elevation

Wing cross-section ▼

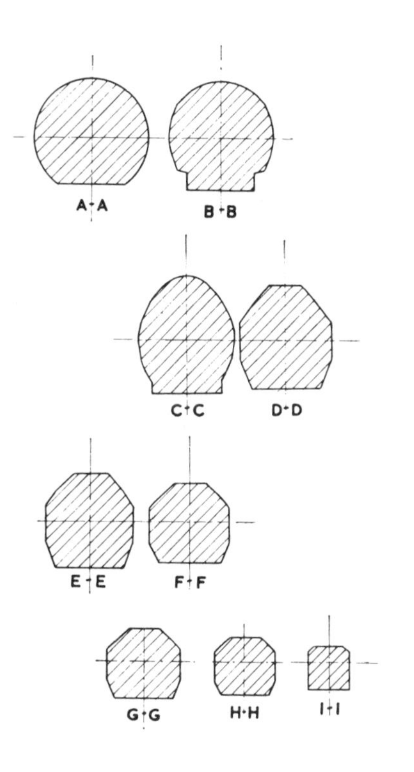

▲ Fuselage cross-sections

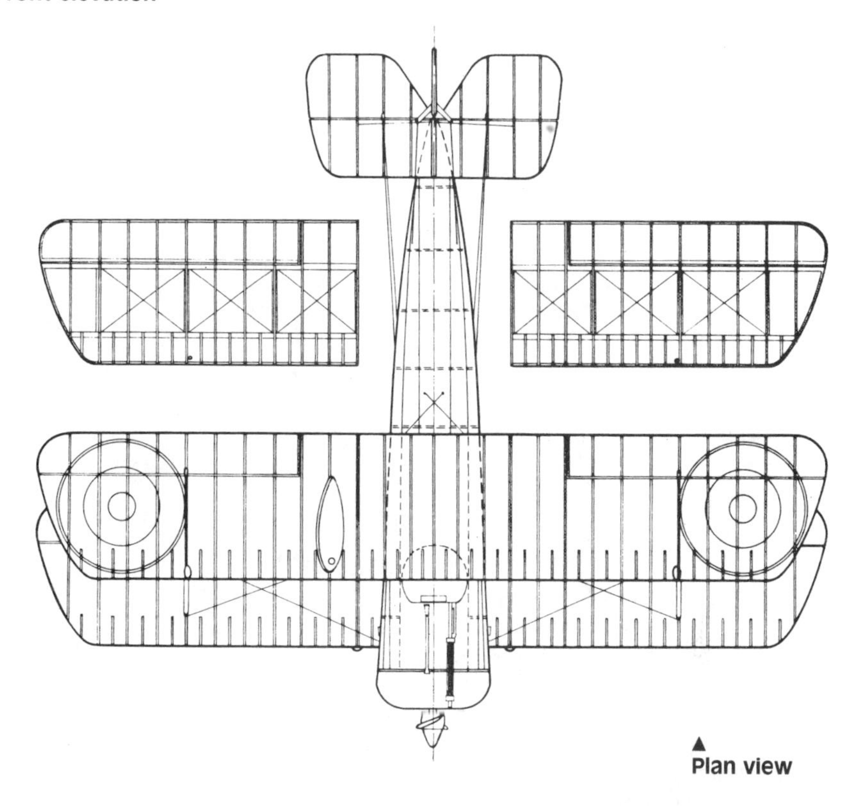

▲ Plan view

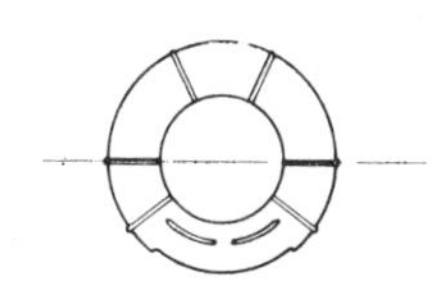

▲ Scrap front elevation

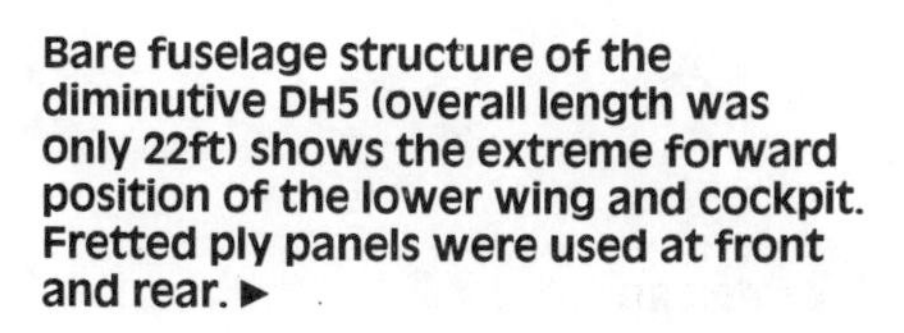

Bare fuselage structure of the diminutive DH5 (overall length was only 22ft) shows the extreme forward position of the lower wing and cockpit. Fretted ply panels were used at front and rear. ▶

Royal Aircraft Factory BE2e

Country of origin: Great Britain.
Type: Two-seat corps reconnaissance aircraft.
Powerplant: One RAF 1a engine rated at 90hp.
Dimensions: Wing span 40ft 9in *12.42m*; length 27ft 3in *8.31m*; height 12ft 0in *3.66m*; wing area 360 sq ft *33.44m²*.
Weights: Empty 1431lb *649kg*; loaded 2100lb *952kg*.
Performance: Maximum speed 82mph *132kph* at 6500ft *1980m*; time to 10,000ft *3050m*, 53min; service ceiling 11,000ft *3350m*; endurance 3.25hr.
Armament: Military load 430lb *195kg*: bombs beneath wings or forward fuselage, plus (typically) one flexibly mounted Lewis machine gun.
Service: Service entry 1916.

DRAWN BY KENNETH McDONOUGH

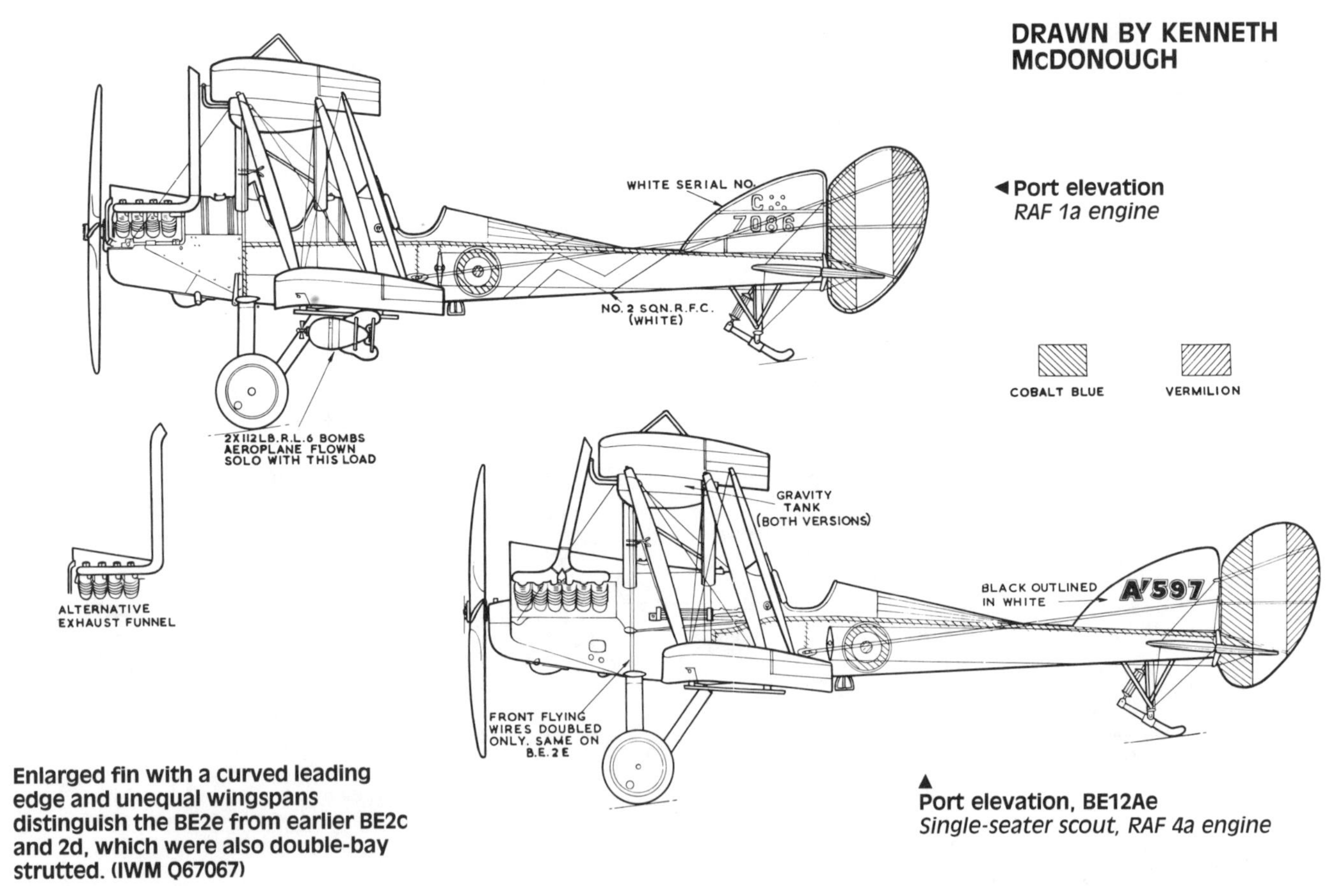

◄ **Port elevation**
RAF 1a engine

▲ **Port elevation, BE12Ae**
Single-seater scout, RAF 4a engine

Enlarged fin with a curved leading edge and unequal wingspans distinguish the BE2e from earlier BE2c and 2d, which were also double-bay strutted. (IWM Q67067)
▼

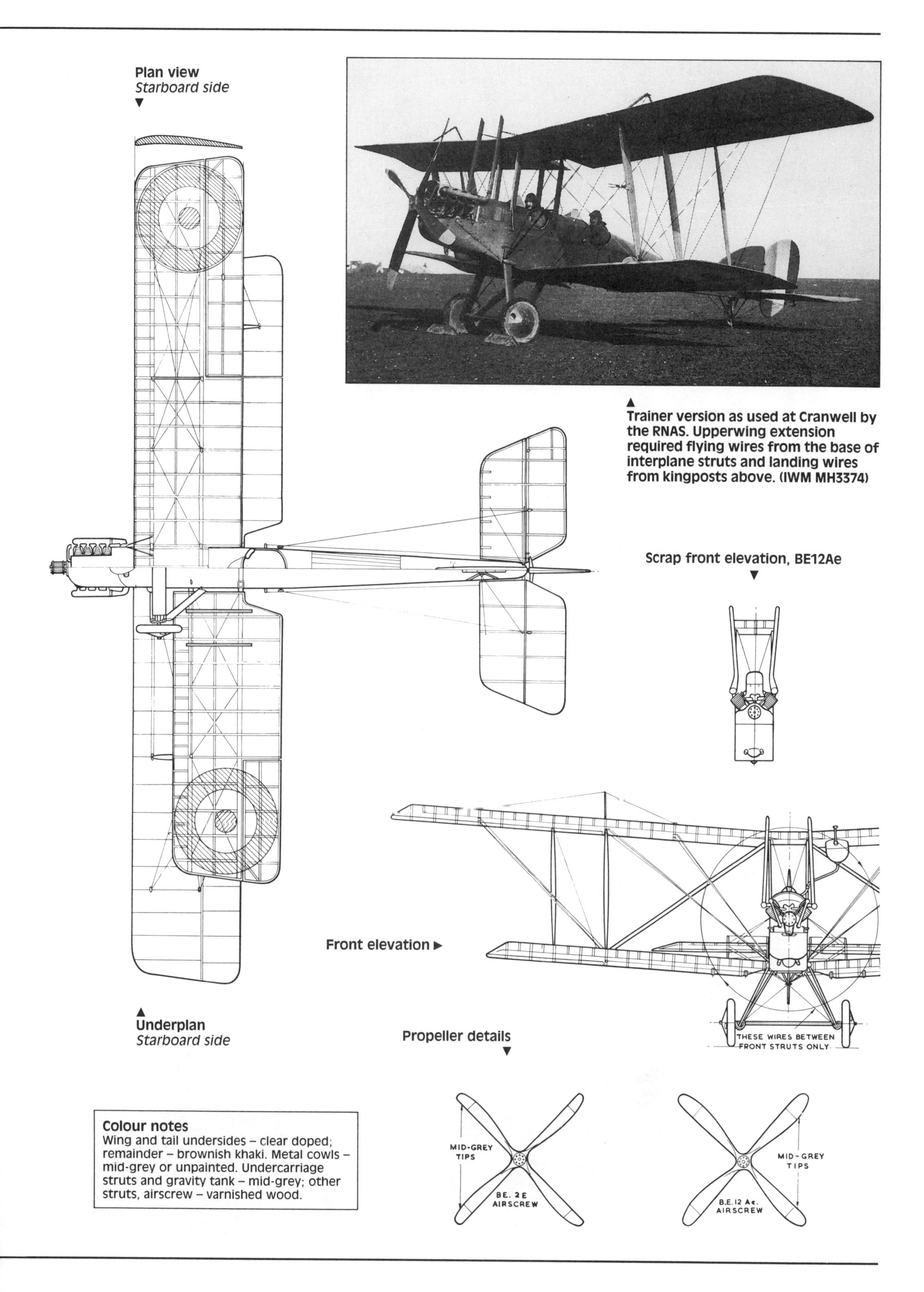

Trainer version as used at Cranwell by the RNAS. Upperwing extension required flying wires from the base of interplane struts and landing wires from kingposts above. (IWM MH3374)

Colour notes
Wing and tail undersides – clear doped; remainder – brownish khaki. Metal cowls – mid-grey or unpainted. Undercarriage struts and gravity tank – mid-grey; other struts, airscrew – varnished wood.

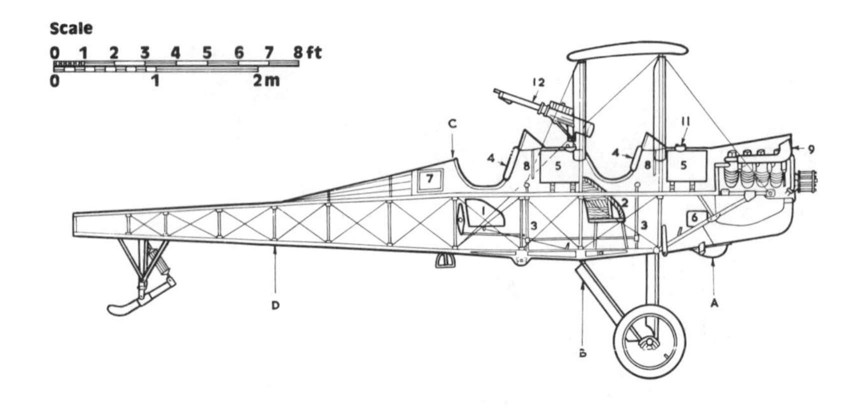

Numerical key
1. Pilot's seat. **2.** Observer's seat. **3.** Control columns. **4.** Leather padding. **5.** Petrol tanks. **6.** Oil tank. **7.** Locker. **8.** Instrument panels. **9.** Alternative exhaust manifold. **10.** Rudder bars. **11.** Filler cap. **12.** Lewis gun (between rear centre-section struts).

Scrap views
Structure
◄▼

10 10 9

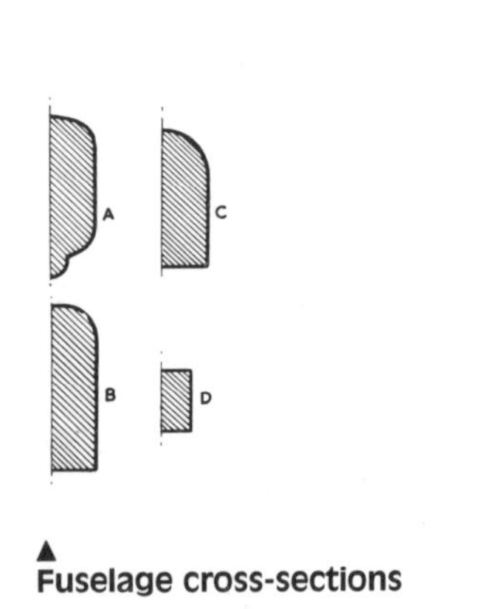

▲
Fuselage cross-sections

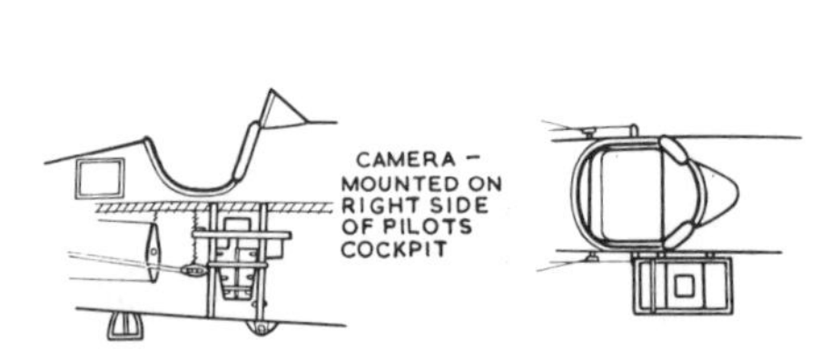

Scrap view
Observer's ammunition drum rack
▼

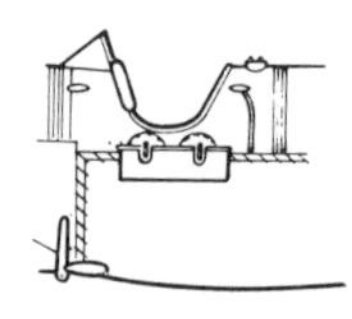

Horn-balanced ailerons of generous proportions appeared on the BE12a, developed from the BE2e wing arrangement, and almost killed Captain L R Tait-Cox on its first flight! (IWM Q57625)
▼

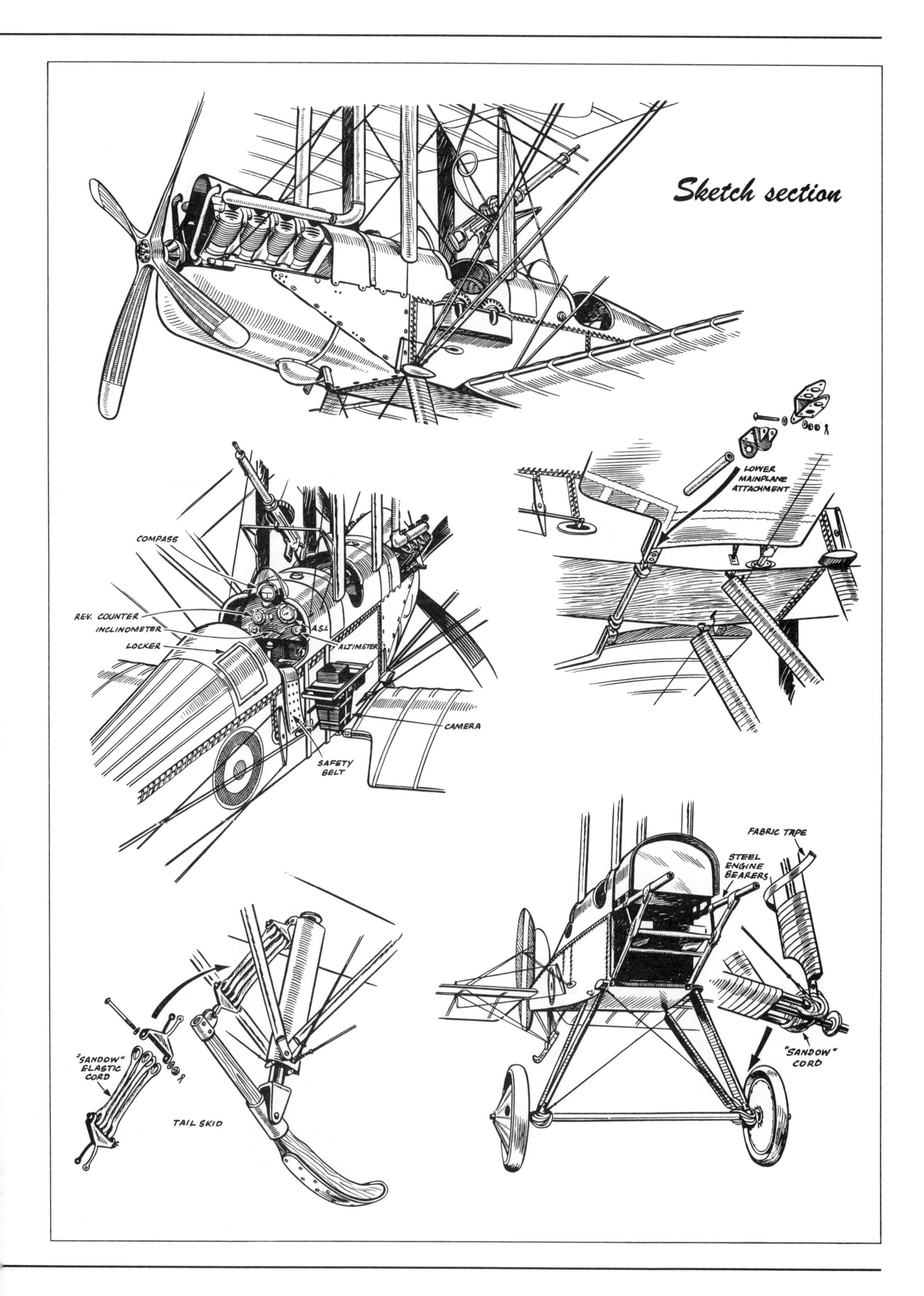
Sketch section
LOWER MAINPLANE ATTACHMENT
COMPASS
REV. COUNTER
INCLINOMETER
LOCKER
A.S.I.
ALTIMETER
CAMERA
SAFETY BELT
FABRIC TAPE
STEEL ENGINE BEARERS
"SANDOW" ELASTIC CORD
TAIL SKID
"SANDOW" CORD

Royal Aircraft Factory RE8

Country of origin: Great Britain.
Type: Two-seat corps reconnaissance aircraft.
Powerplant: One RAF 4a twelve-cylinder vee engine rated at 150hp.
Dimensions: Wing span 42ft 7in *12.98m*; length 27ft 10in *8.48m*; height 11ft 4in *3.45m*; wing area 377.5 sq ft *35.1m²*.
Weights: Empty 1803lb *818kg*; loaded 4109lb *1252kg*.
Performance: Maximum speed 102mph *164kph* at 6500ft *1980m*; time to 6500ft, 15min; service ceiling 13,000ft *3960m*; endurance 4.5hr.
Armament: One fixed Vickers machine gun and one single or twin flexibly mounted Lewis machine gun.
Service: Service entry November 1916.

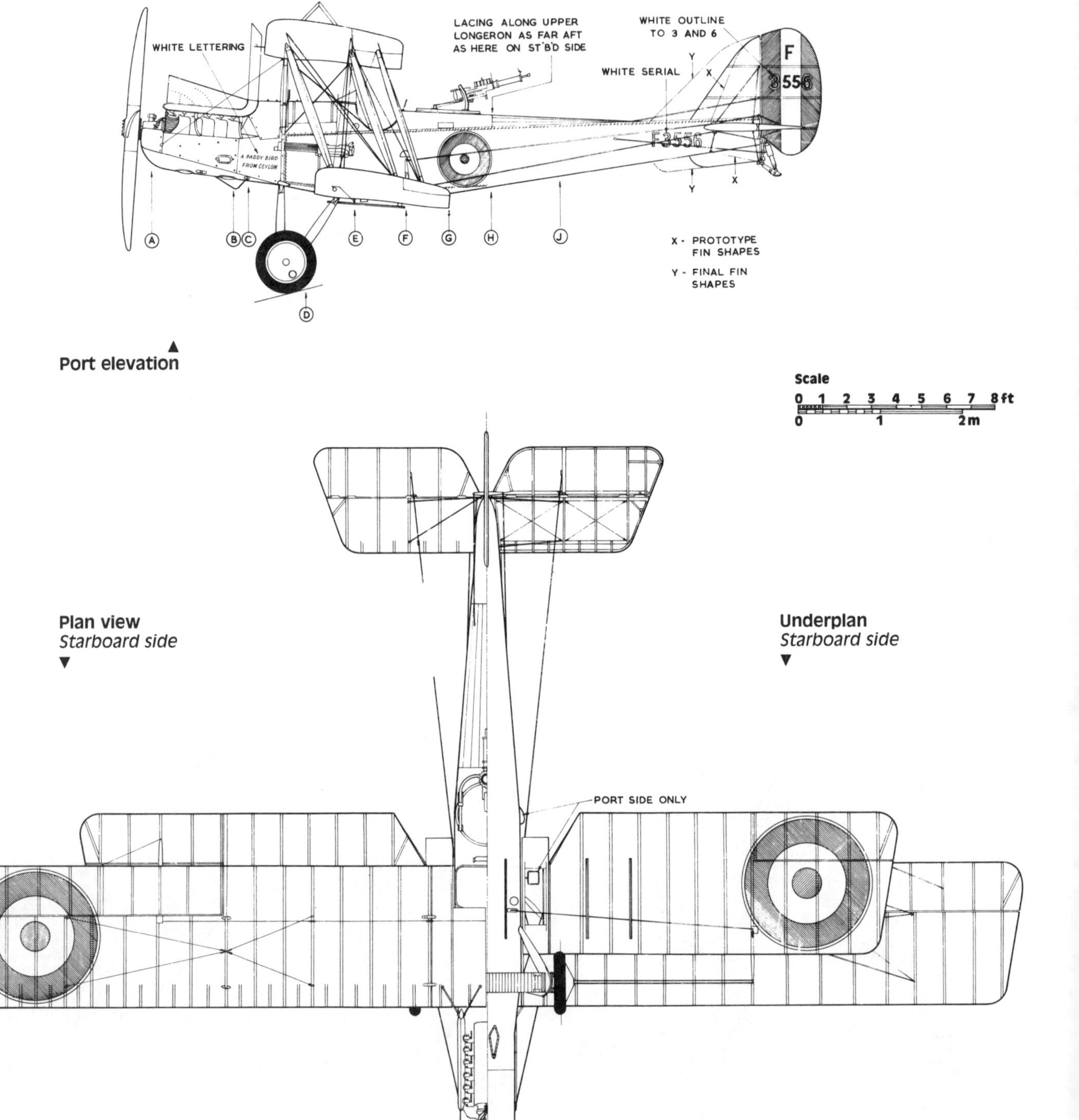

Port elevation

Scale

Plan view
Starboard side

Underplan
Starboard side

DRAWN BY G A G COX

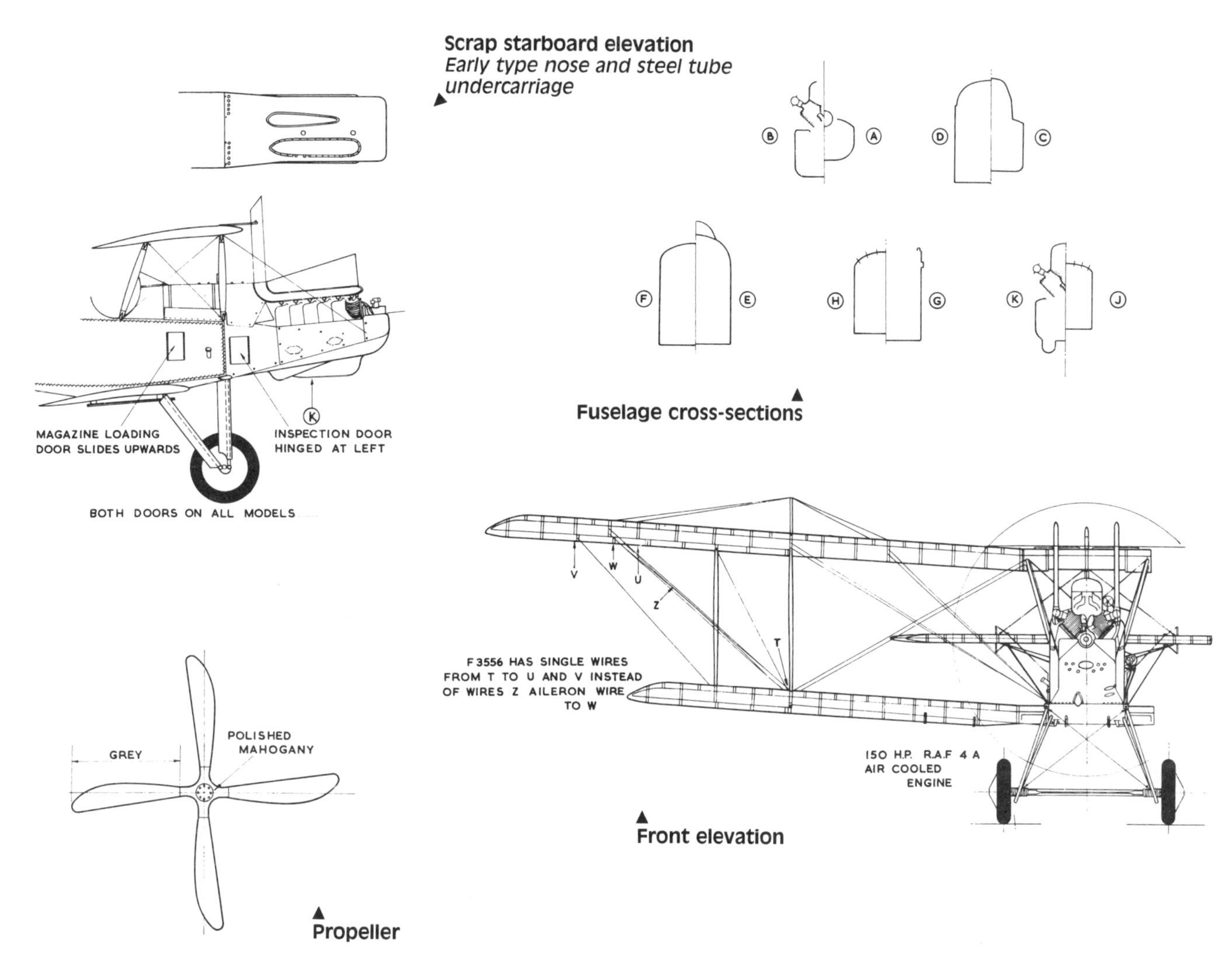

Scrap starboard elevation
Early type nose and steel tube undercarriage

Fuselage cross-sections

Front elevation

Propeller

Prototype RE8 with small fin and pillar-type gun mount. Finish is varnished Irish linen. Flying surfaces had similar shapes and structures to those of the BE2e. (IWM 66027)

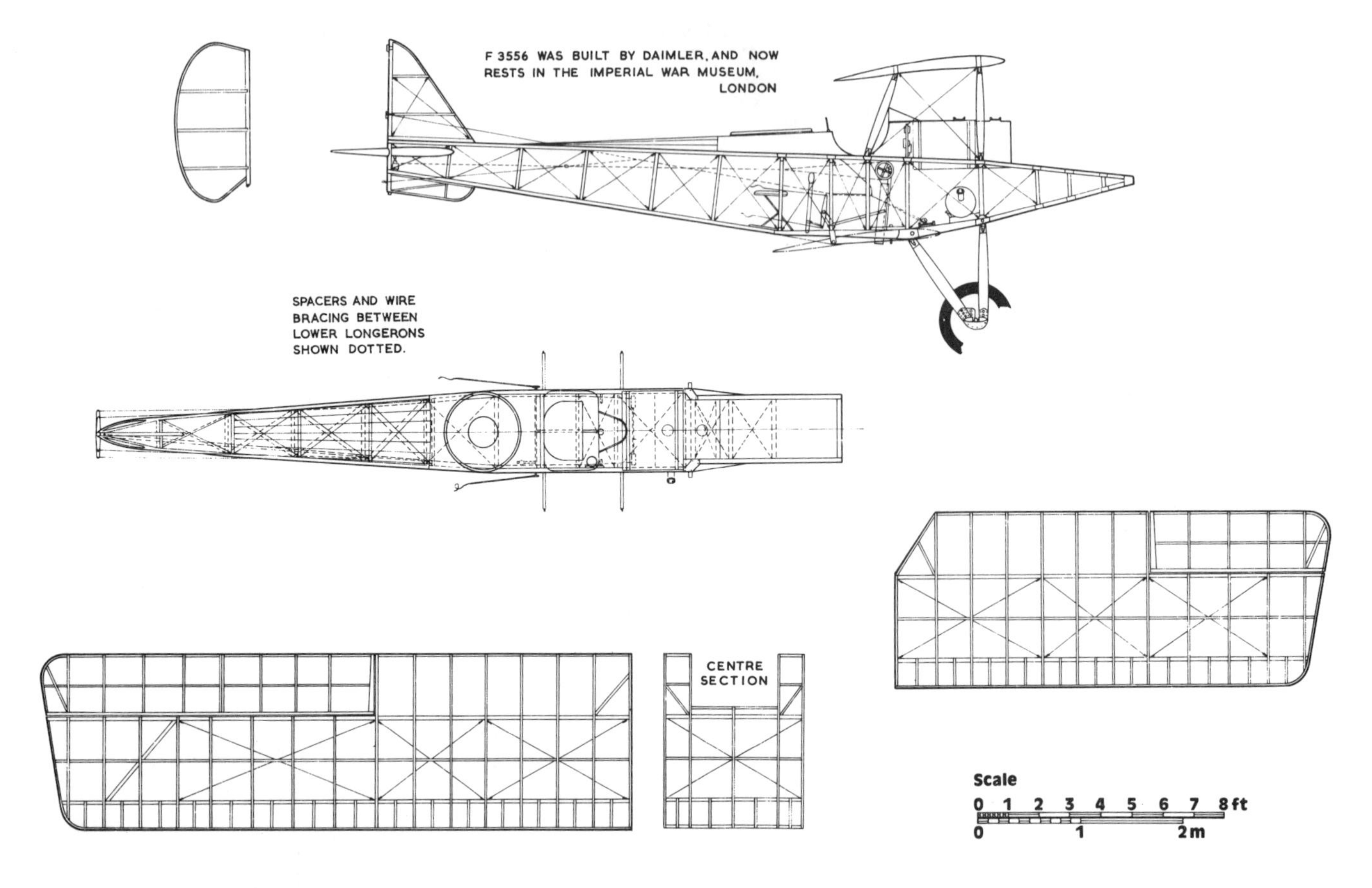

▲
Scrap views
Structure

Standard production RE8 with solid ash undercarriage struts and enlarged under fin, applied to make 'Harry Tate' more manageable. (Air Min H1287)
▼

Sketch section

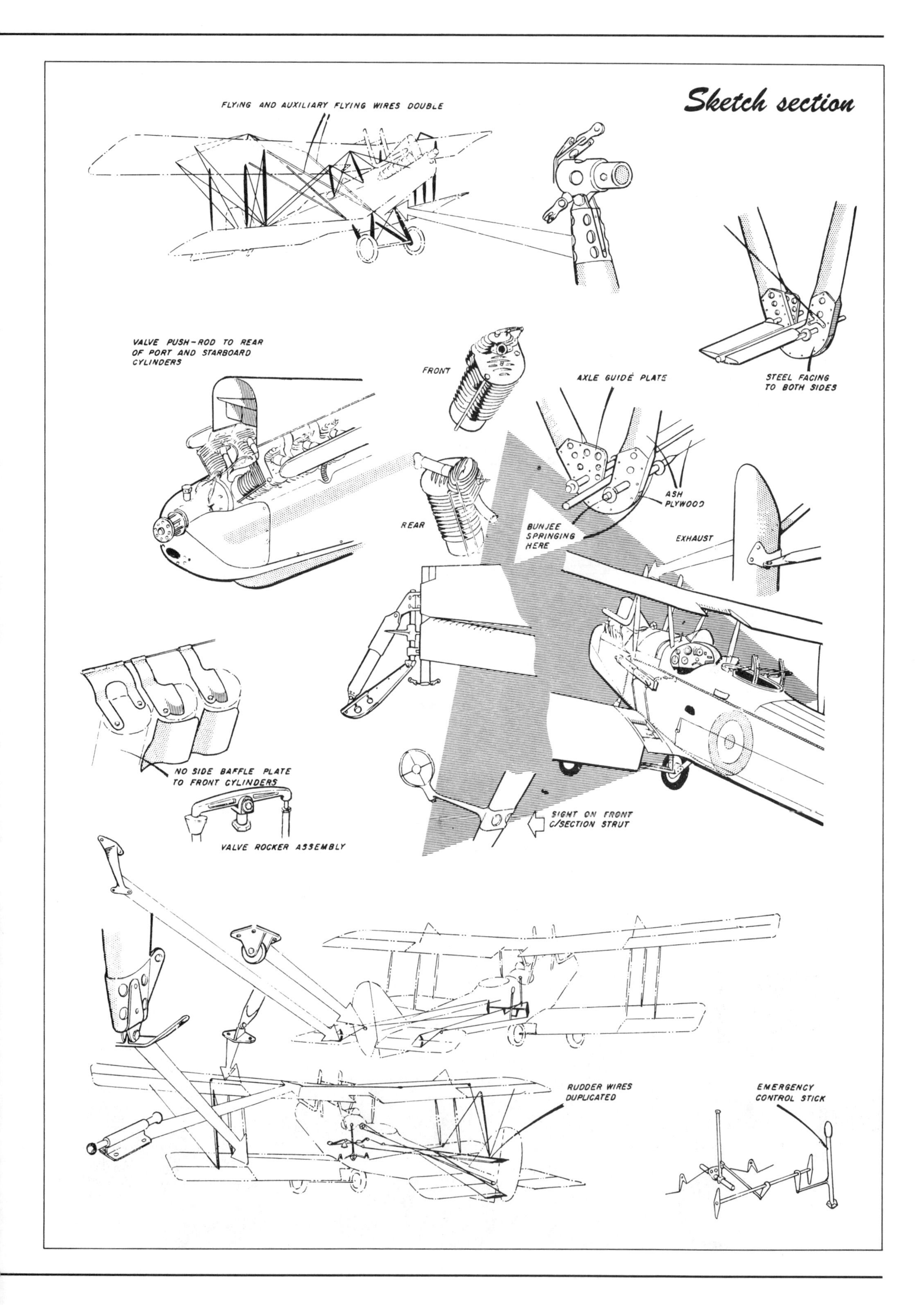

Sopwith 5F1 Dolphin

Country of origin: Great Britain.
Type: Single-seat fighter.
Powerplant: One Hispano V8 engine rated at 200hp.
Dimensions: Wing span 32ft 6in *9.91m*; length 22ft 3in *6.78m*; height 8ft 6in *2.59m*; wing area 263.25 sq ft *24.46m²*.
Weights: Loaded 1959lb *888kg*.
Performance: (Loaded) Maximum speed 121.5mph *196kph* at 10,000ft *3050m*; time to 10,000ft, 12.1min; service ceiling 20,000ft *7100m*.
Armament: Two fixed Vickers machine guns and (some aircraft) one or two Lewis machine guns.
Service: First flight May 1917; service entry late 1917.

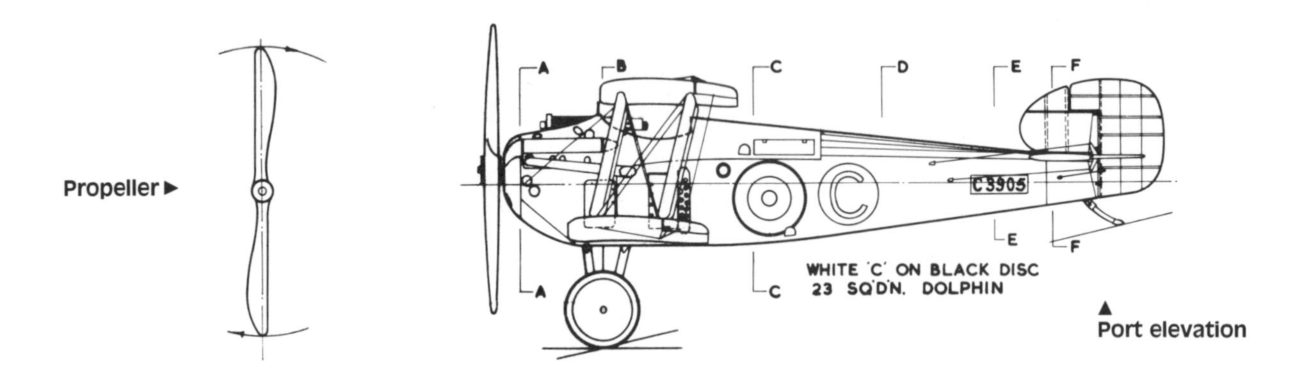

Propeller ▶

▲ Port elevation

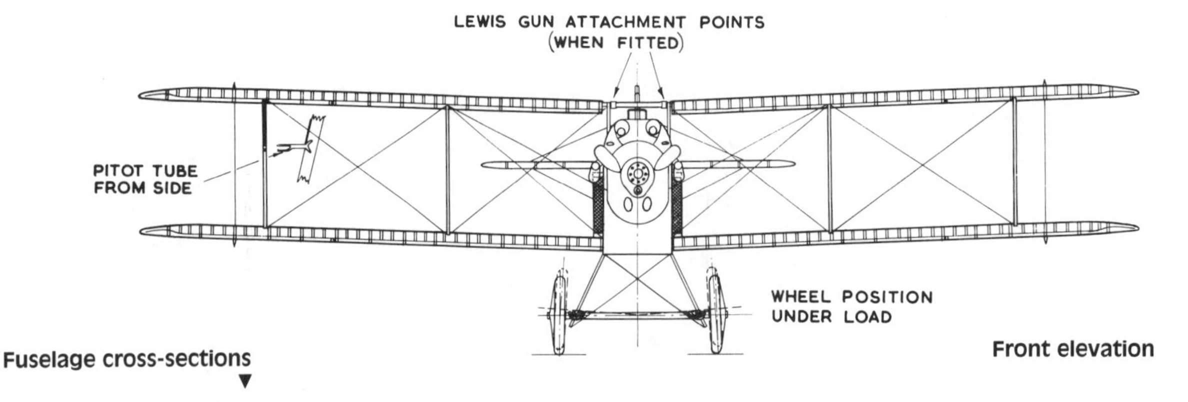

Front elevation

Fuselage cross-sections ▼

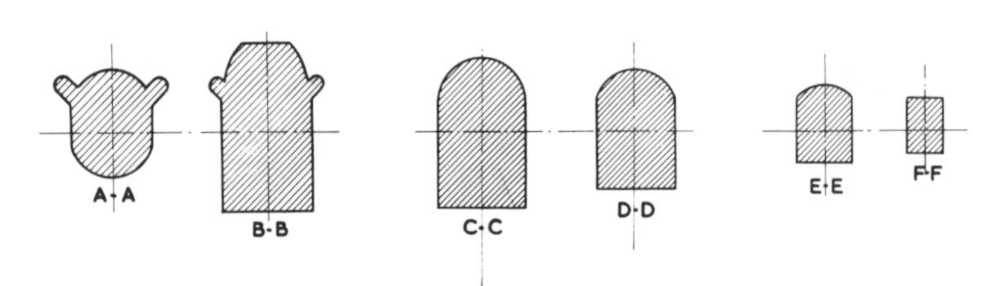

The pilot of this Dolphin (which was adapted for night flying), was protected by half-loops of steel tubing above the inner interplane struts, to act as roll-over supports in case of landing accidents. Fear of decapitation led to the name 'Blockbuster'. (IWM Q66095) ▼

▲ Twin upward-firing gun mount is seen on this Dolphin, the production version with side radiators for the 200hp Hispano-Suiza engine. Designed to give the pilot the best possible view, the type was not as hazardous as anticipated and saw action with Nos 19, 87 and 98 Sqns. (IWM Q67512)

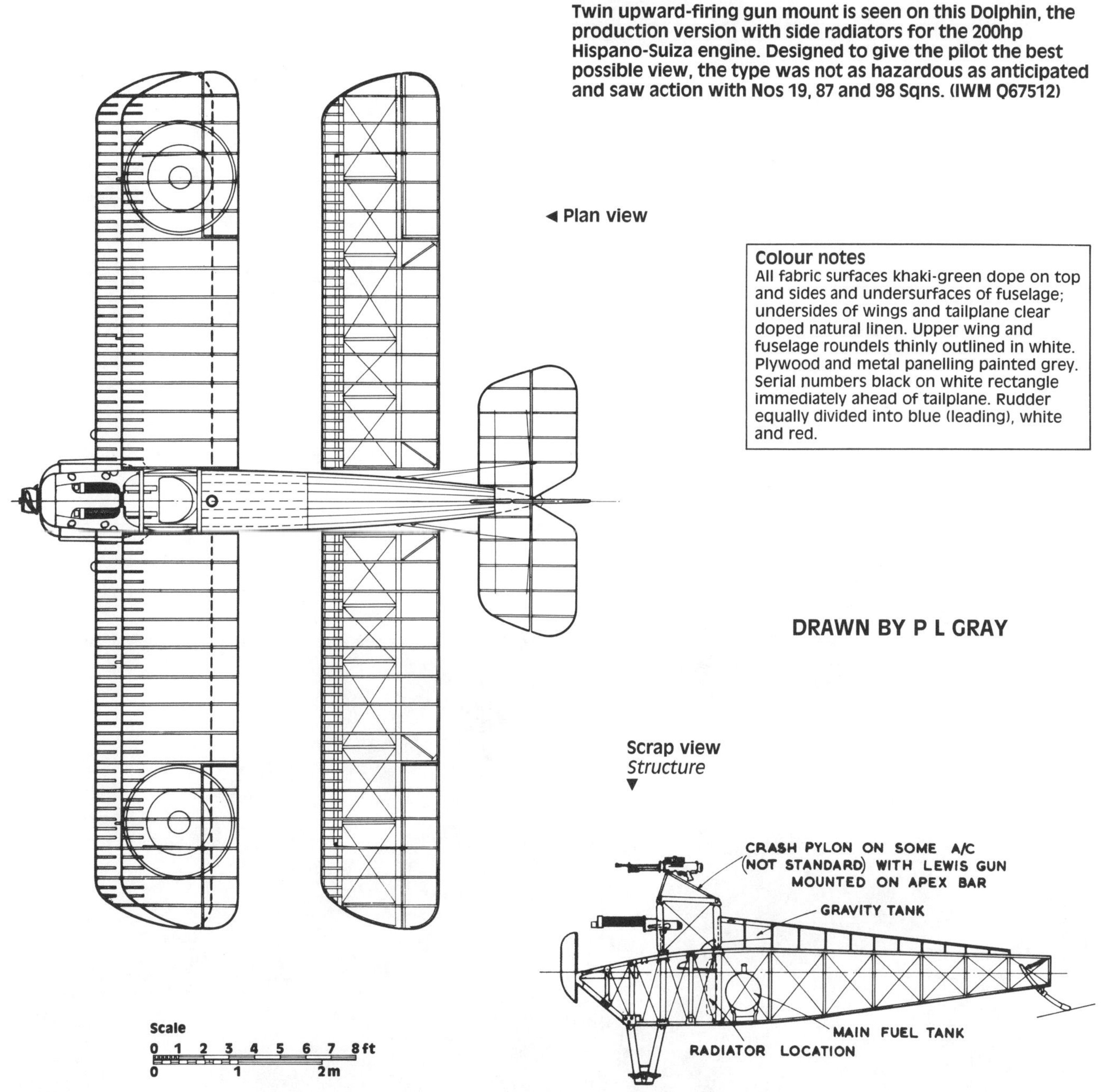

Colour notes
All fabric surfaces khaki-green dope on top and sides and undersurfaces of fuselage; undersides of wings and tailplane clear doped natural linen. Upper wing and fuselage roundels thinly outlined in white. Plywood and metal panelling painted grey. Serial numbers black on white rectangle immediately ahead of tailplane. Rudder equally divided into blue (leading), white and red.

Sopwith Pup

Country of origin: Great Britain.
Type: Single-seat fighter.
Powerplant: One Le Rhône or Clerget engine rated at 80hp or Gnome rated at 100hp.
Dimensions: Wing span 26ft 6in *8.08m*; length 19ft 3½in *5.88m*; height 9ft 5in *2.87m*; wing area 276 sq ft *25.64m*.
Weights: Empty (Le Rhône) 787lb *357kg*, (Gnome) 856lb *388kg*; loaded (Le Rhône) 1225lb *556kg*, (Gnome) 1297lb *588kg*.
Performance: Maximum speed (Le Rhône) 111.5mph *179.5kph*, (Gnome) 110mph *177kph*; time to 10,000ft *3050m*, (Le Rhône) 14.25min, (Gnome) 12.25min; endurance (Le Rhône) 3hr, (Gnome) 1.75hr.
Armament: One fixed Vickers machine gun.
Service: First flight February 1916; service entry autumn 1916.

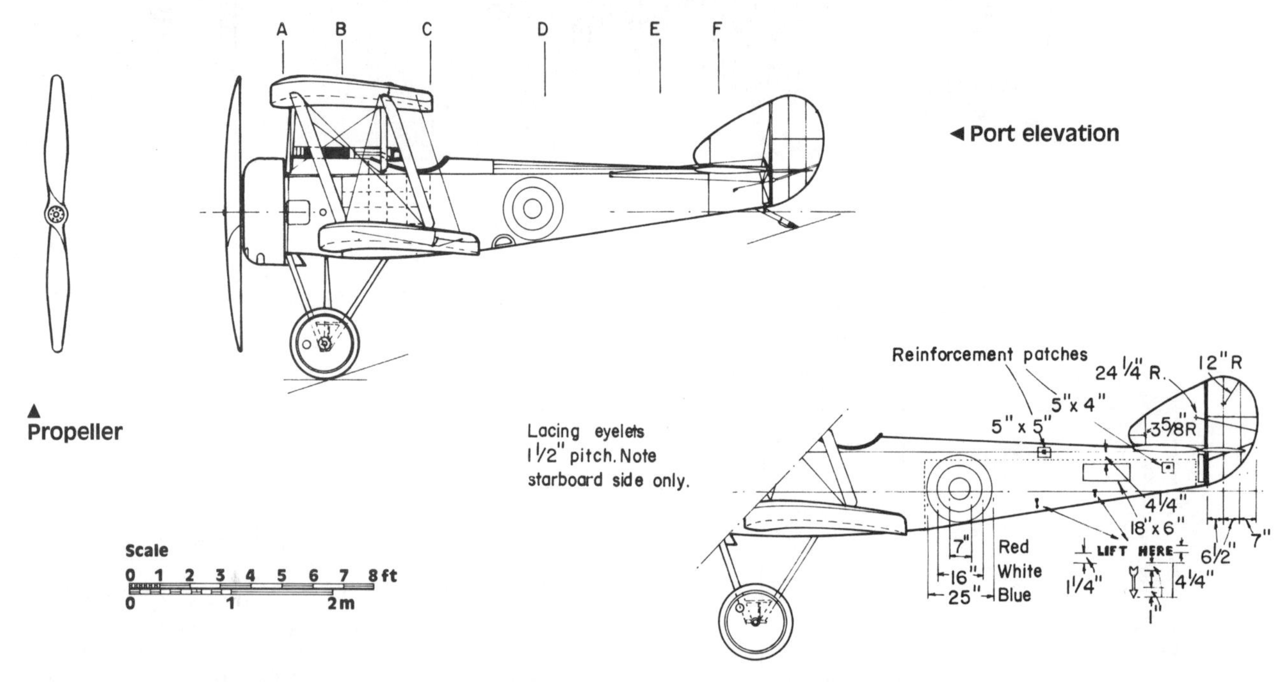

◄ Port elevation

▲ Propeller

▲ Scrap port elevation
Showing original Sopwith dimensions

Pups were fitted with the 80hp Le Rhône or Gnome engine or the 100hp Gnome Monosoupape, as on this machine and distinguished by the cutaway lower cowl; see also rear three-quarter view of B1755.
▼

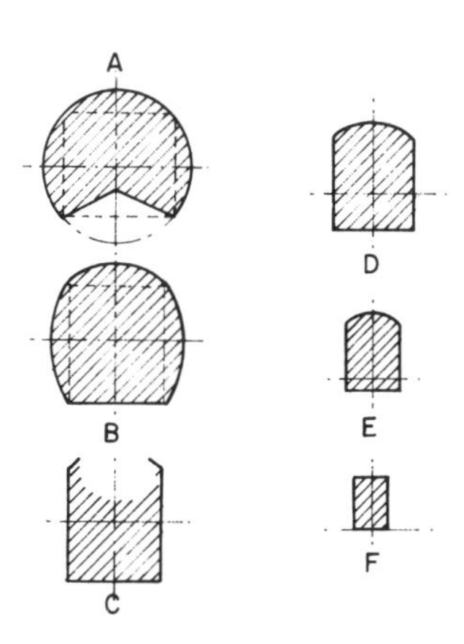

▲
Fuselage cross-sections

▲
Stripes on the elevators of this RNAS Pup in the Middle East Theatre at 3 (Naval) Wing, Imbros, were an additional identification aid. A Lewis gun has been mounted on the centre-section struts to fire above the propeller arc. (J M Bruce/G S Leslie)

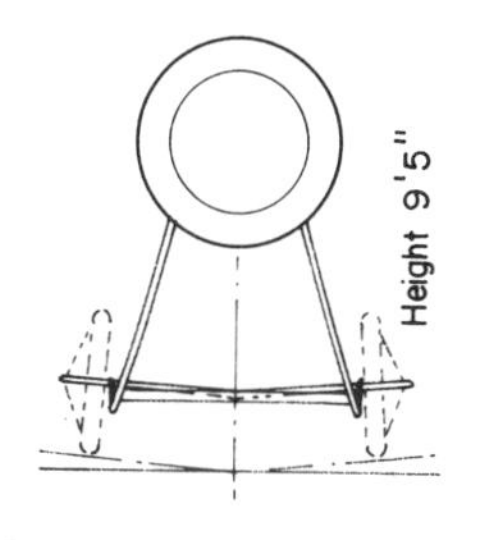

▲
Scrap front elevation
Axle under load

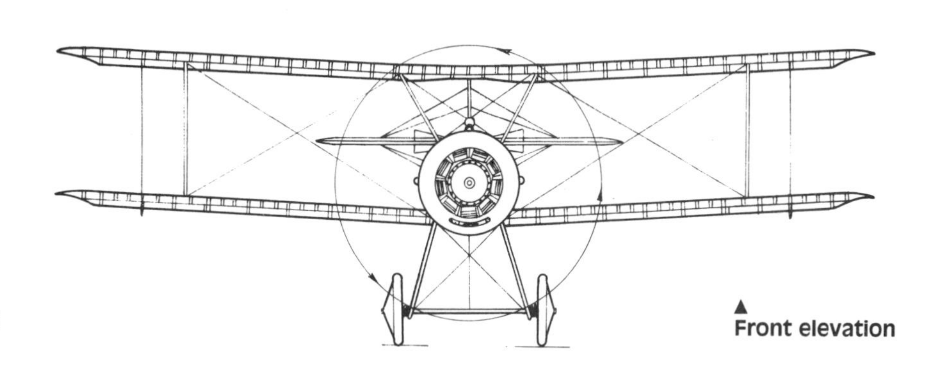

▲
Front elevation

The Standard Motor Co was one of several contractors making the Pup, including B1755 seen here awaiting delivery in its PC10 finish. Favoured by pilots for its manoeuvrability, the Pup subsequently became an excellent trainer.
▼

DRAWN BY P L GRAY

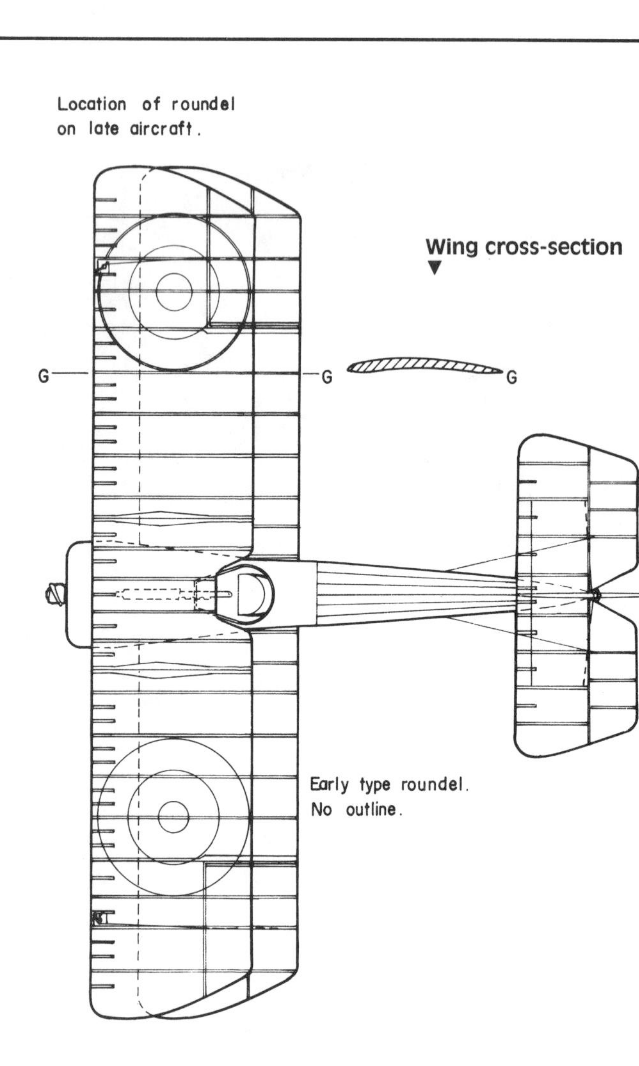

▼ Wing cross-section

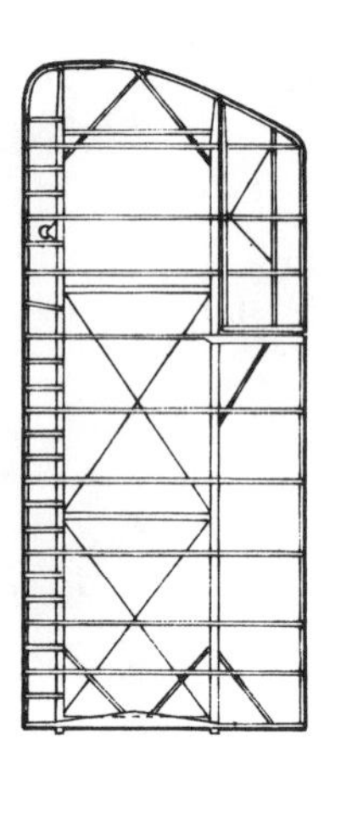

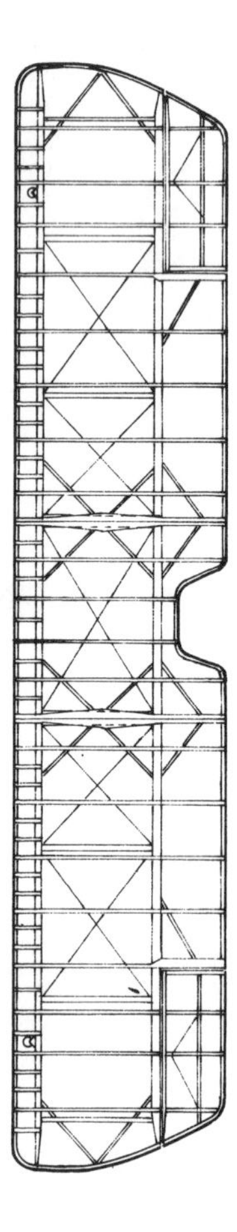

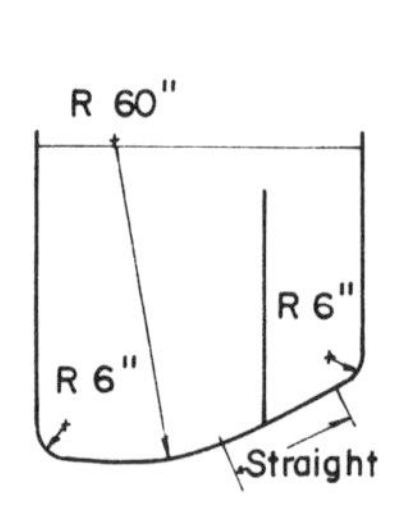

▲ Plan view

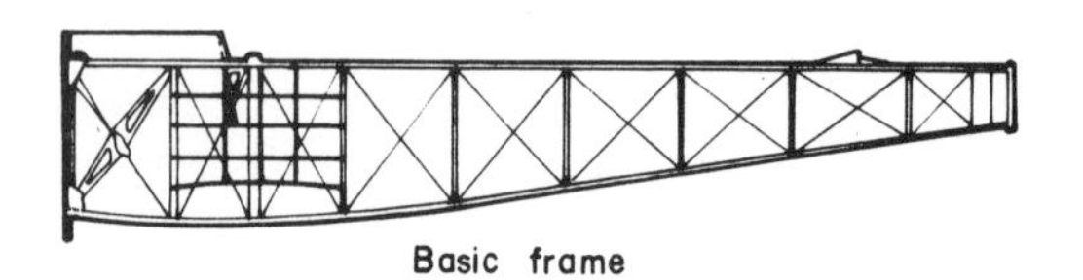

◄▲ Scrap views
Structure

Scale
0 1 2 3 4 5 6 7 8ft
0 1 2m

A Whitehead-built machine, made at Richmond, Surrey, A6231 is readied for flight. The Pup could reach 17,500ft and had an endurance of 3 hours. (IWM Q56011)
▼

Colour notes
Top and side surfaces standard khaki-green with coat of protective varnish; undersurfaces left natural unbleached doped linen, which when varnished resulted in a darkish creamy shade. Cowling and side panels often bright metal, but sometimes overpainted dark grey or khaki-green. Centre-section and undercarriage struts likewise painted, but interplane struts left varnished wood. Roundels at all four wing tips and on fuselage sides, those on top and side surfaces narrowly outlined in white but not invariably so. Shade of khaki-green used on Admiralty aircraft was darker and greener than on Army machines, and fuselage sides of some Admiralty Pups were left in linen finish. Elevators of Navy machines were painted red, white and blue in equal spanwise divisions (blue foremost), as were rudders of all aircraft.

A trademark of the prototype Pup was its fin, with the Company name signwritten in black over the white fabric. Its name is said to have originated with pilots who thought it to be an offspring of the 1½ Strutter. (IWM Q67505) ►

A Royal Flying Corps Pup, built by Whitehead Aircraft Ltd, with a white fuselage band and standard single Vickers gun in front of the cockpit. The gun had Sopwith-Kraupner mechanical synchronising gear. (J M Bruce/ G S Leslie) ►

Rebuilt by No. 2 (Northern) Aircraft Depot, Catterick, this colourful Pup has yellow or orange bands alternating with black around the fuselage and does not appear to be armed. A total of 1770 Pups were built and they served on all fronts in the Great War. (J M Bruce/G S Leslie) ►

Sopwith Pup fitted with a skid chassis, most probably an RNAS machine as it has elevator stripes and would be used for deck-landing experiments. This example is in natural varnish finish. (J M Bruce/G S Leslie) ►

Sopwith 7F1 Snipe

Country of origin: Great Britain.
Type: Single-seat fighter.
Powerplant: One Bentley BR2 rotary engine rated at 228hp.
Dimensions: Wing span 31ft 1in *9.47m*; length 19ft 10in *6.05m*.
Weights: Empty 1312lb *595kg*; loaded 2020lb *916kg*.
Performance: Maximum speed 121mph *195kph* at 10,000ft *3050m*; time to 10,000ft, 9.4min; service ceiling 19,500ft *5945m*; endurance 3hr.
Armament: Two fixed Vickers machine guns.
Service: First flight December 1917.

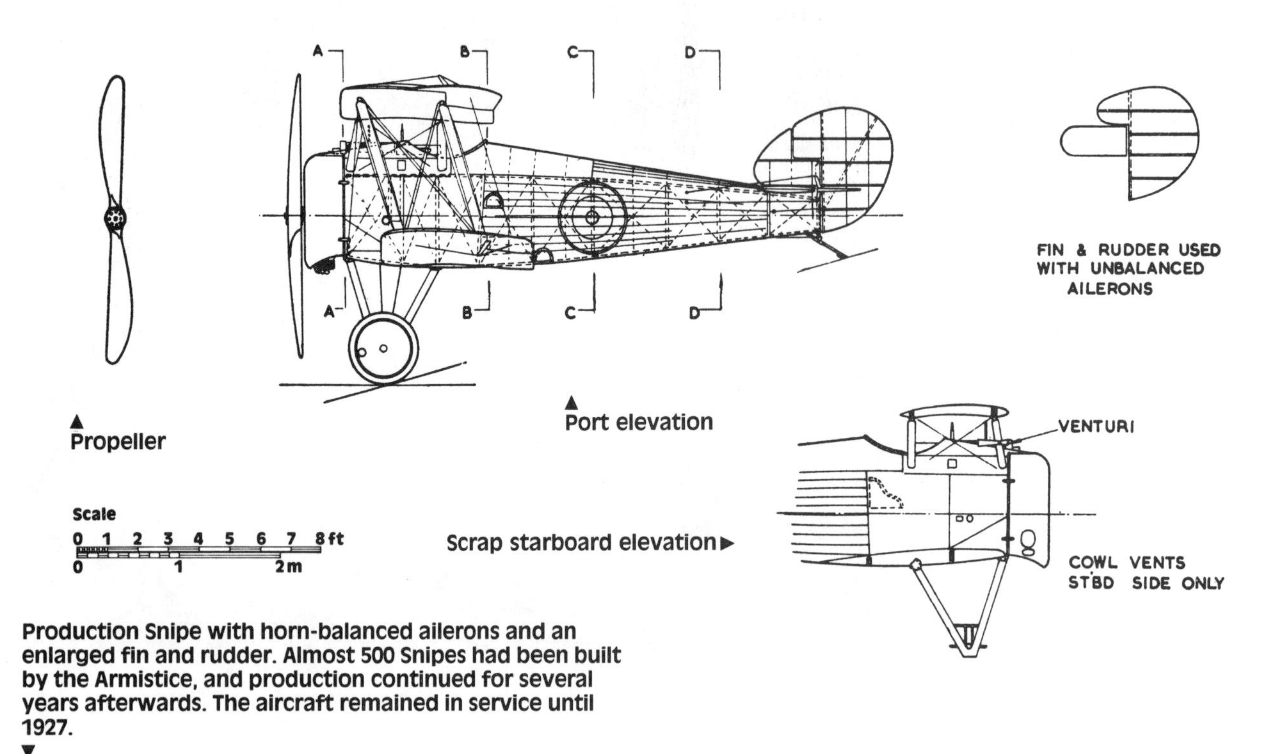

▲ Propeller

▲ Port elevation

Scrap starboard elevation ▶

Production Snipe with horn-balanced ailerons and an enlarged fin and rudder. Almost 500 Snipes had been built by the Armistice, and production continued for several years afterwards. The aircraft remained in service until 1927.
▼

Colour notes
All fabric top and side surfaces standard khaki-green, undersurfaces clear doped and varnished linen. Ply panelling medium grey, with struts either varnished natural spruce or fabric-covered and doped khaki-green. Serial numbers introduced August 1918 on rudder and rear fuselage: ex-works machines had fuselage serials in black digits on an oblong white background, with rudder serials appearing fraction fashion (i.e. letter above numerals), superimposed on rudder stripes and outlined white. Variations occurred on operational aircraft. Postwar Snipes eventually finished in aluminium dope, with black serial numbers under wings; prefix letter was half the height of the numerals. Aircraft illustrated is in 4 Australian FC Sqn markings, stationed at Bickendorf shortly after the Armistice.

▲ Night fighting Snipe with red/blue roundels, white edged as if to ensure that the markings are visible!

DRAWN BY P L GRAY

Front elevation ▼

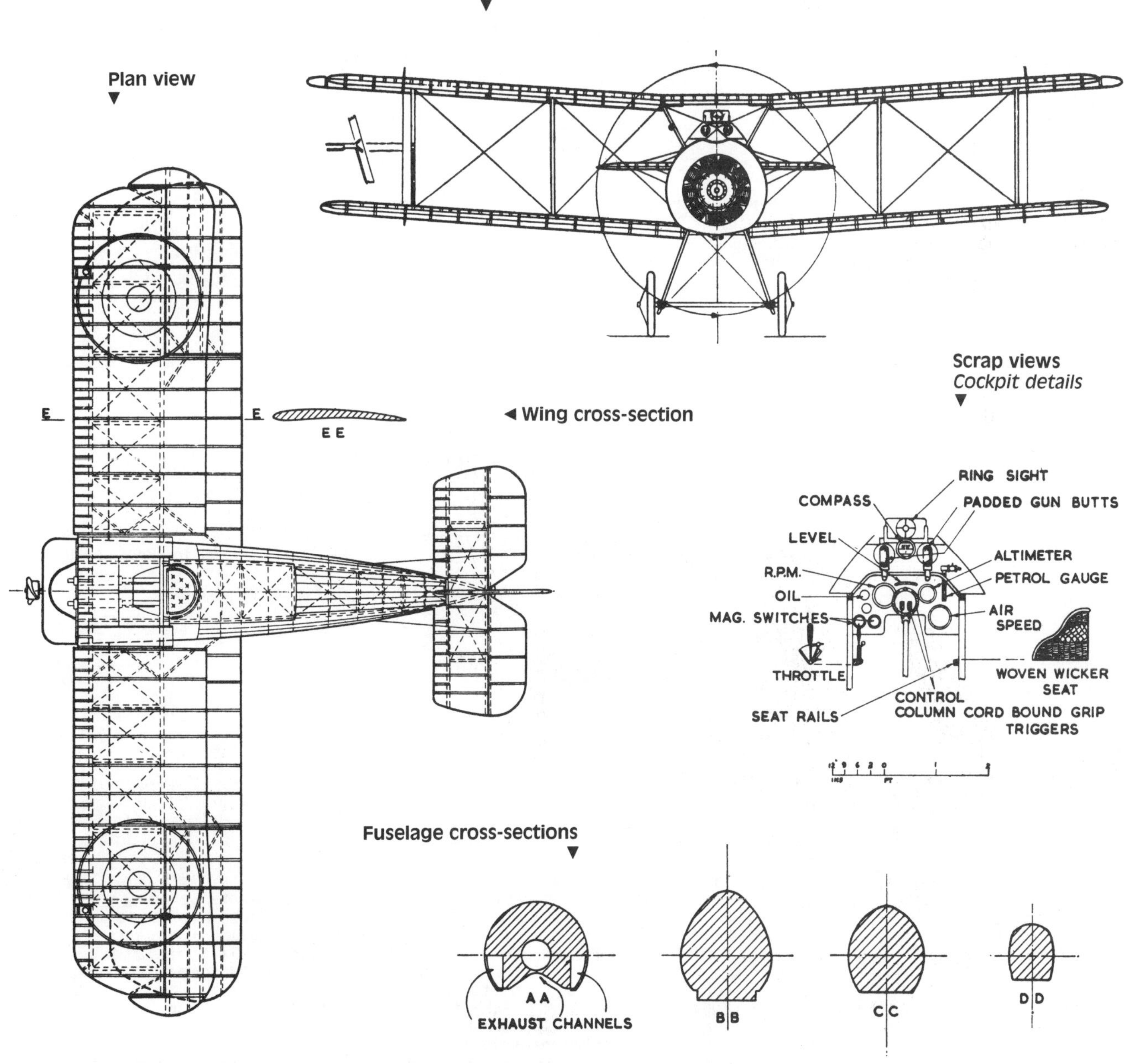

Sopwith Triplane

Country of origin: Great Britain.
Type: Single-seat fighter.
Powerplant: One Clerget rotary engine rated at 110hp (later 130hp).
Dimensions: Wing span 26ft 6in *8.08m*; length 18ft 10in *5.74m*.
Weights: Empty 993lb *450kg*; loaded 1415lb *642kg*.
Performance: Maximum speed 116mph *187kph* at 6500ft *1980m*; time to 10,000ft *3050m*, 10.5min.
Armament: One fixed Vickers machine gun.
Service: First flight May 1916; service entry early 1917.

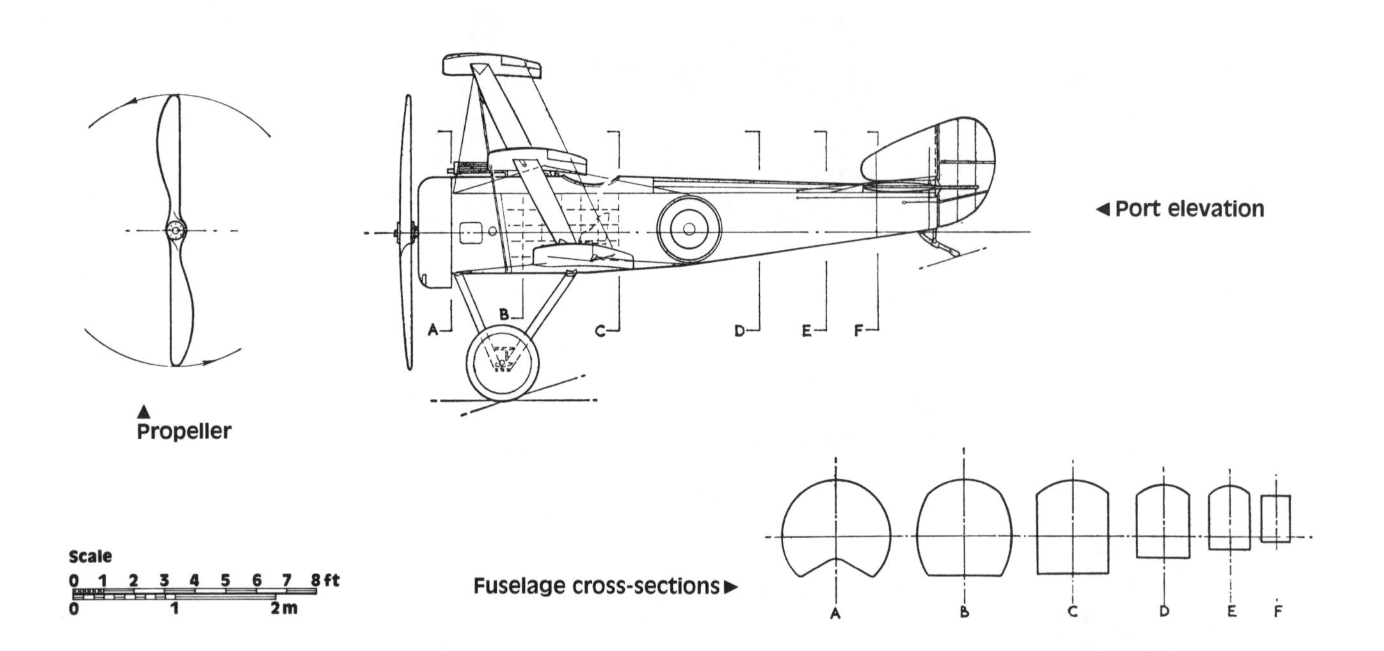

◄ Port elevation

▲ Propeller

Fuselage cross-sections ►

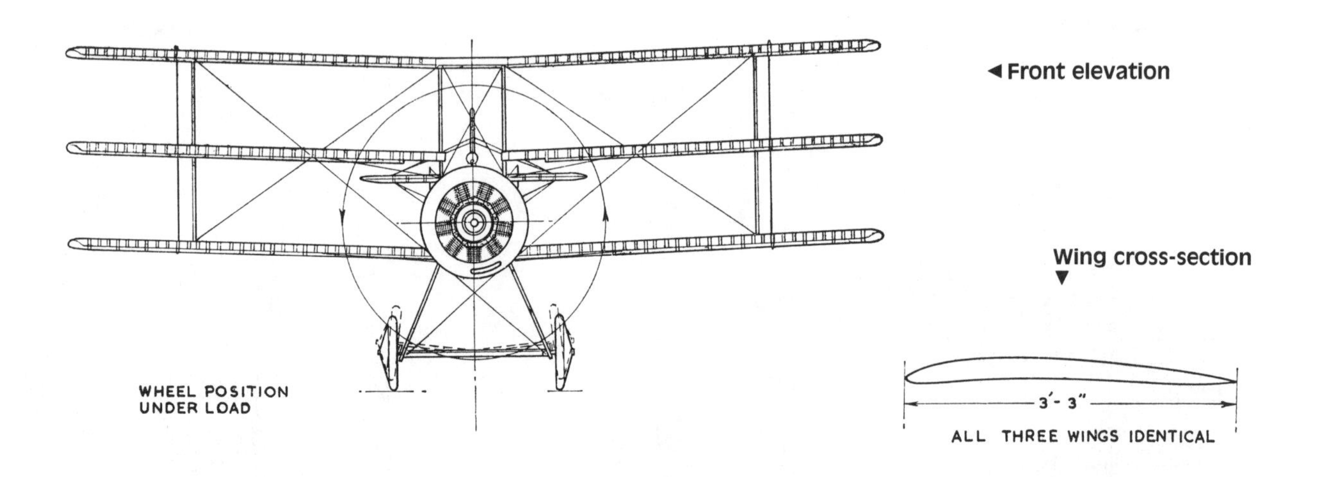

◄ Front elevation

Wing cross-section ▼

Colour notes

Except for prototype, which was left plain, unbleached fabric overall, aircraft were doped in regulation khaki-green on upper surfaces and on vertical surfaces of fuselage; undersides were plain doped fabric, darkening when varnished. Cowling and metal panels often left bright by factories, but invariably overpainted dark grey or khaki-green on operations. Wheel discs were left natural fabric by makers, but again were often doped khaki-green in usage, as was also the case with the fins of Sopwith-built machines. Serial numbers were painted on rear fuselage just ahead of tailplane, either in white or in black superimposed on a small white rectangle. Roundels appeared above and below wings (full chord) and on fuselage sides, those against a khaki background being narrowly outlined in white (not always for fuselage roundels). Blue (forward), white and red rudder stripes were of equal width. Triplanes of No 1 (Naval) Sqn dispensed with fuselage roundels and bore a rectangular white/red/white flash ahead of the serial; between this marking and the cockpit, extra large white numerals, the full width between the longerons, were carried – 1 to 18.

▲
One of a batch of 75 Triplanes built by the Sopwith Aviation Co, N5430 was transferred from the RNAS to the RFC for evaluation.

No 1 Sqn RNAS, headed by Flt Cdr H V Rowley, whose machine (7) has a white fin. Fuselage roundels were not carried by this unit, famous for many aerial battles in 1917. (IWM Q66794)
▼

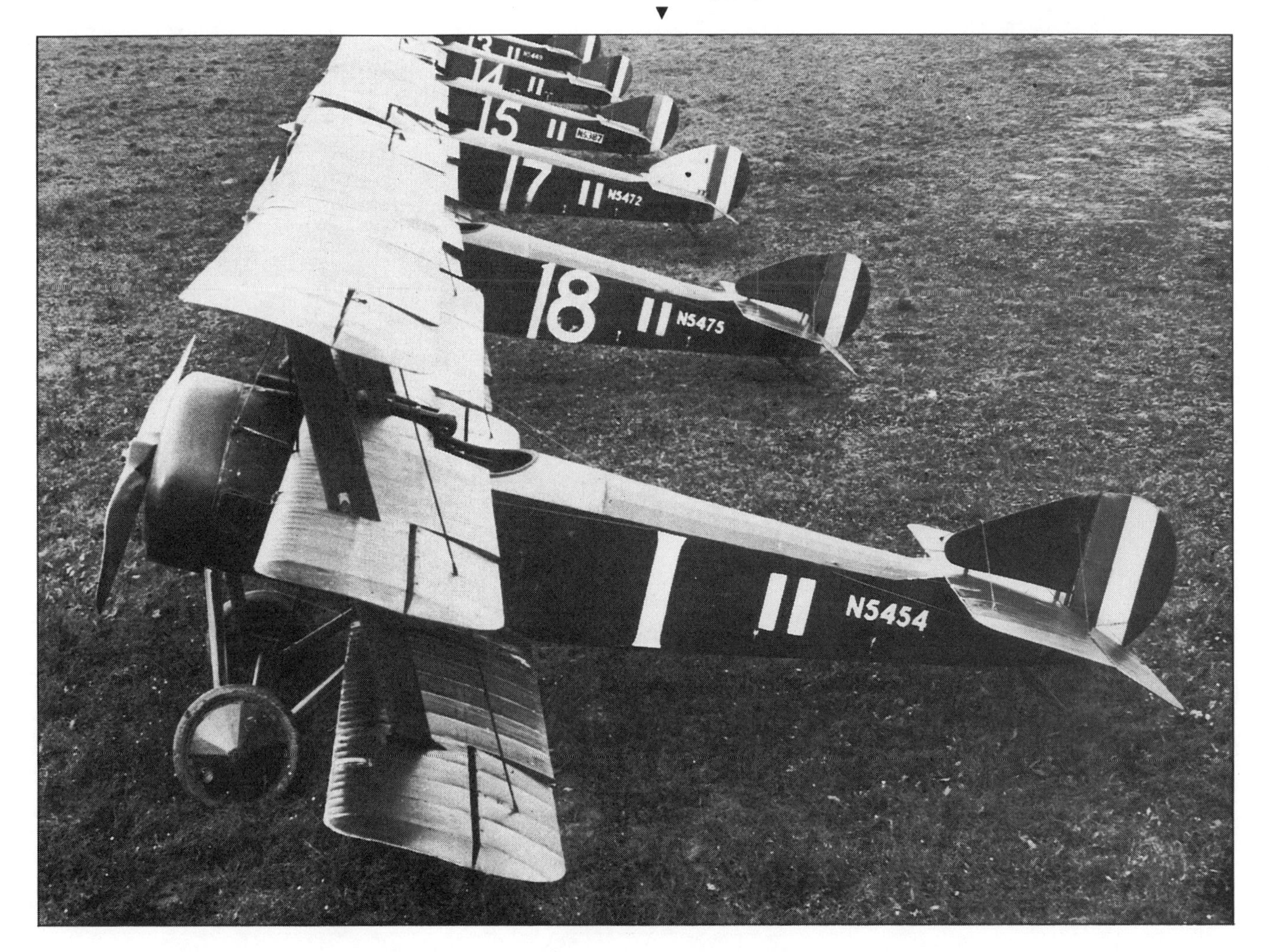

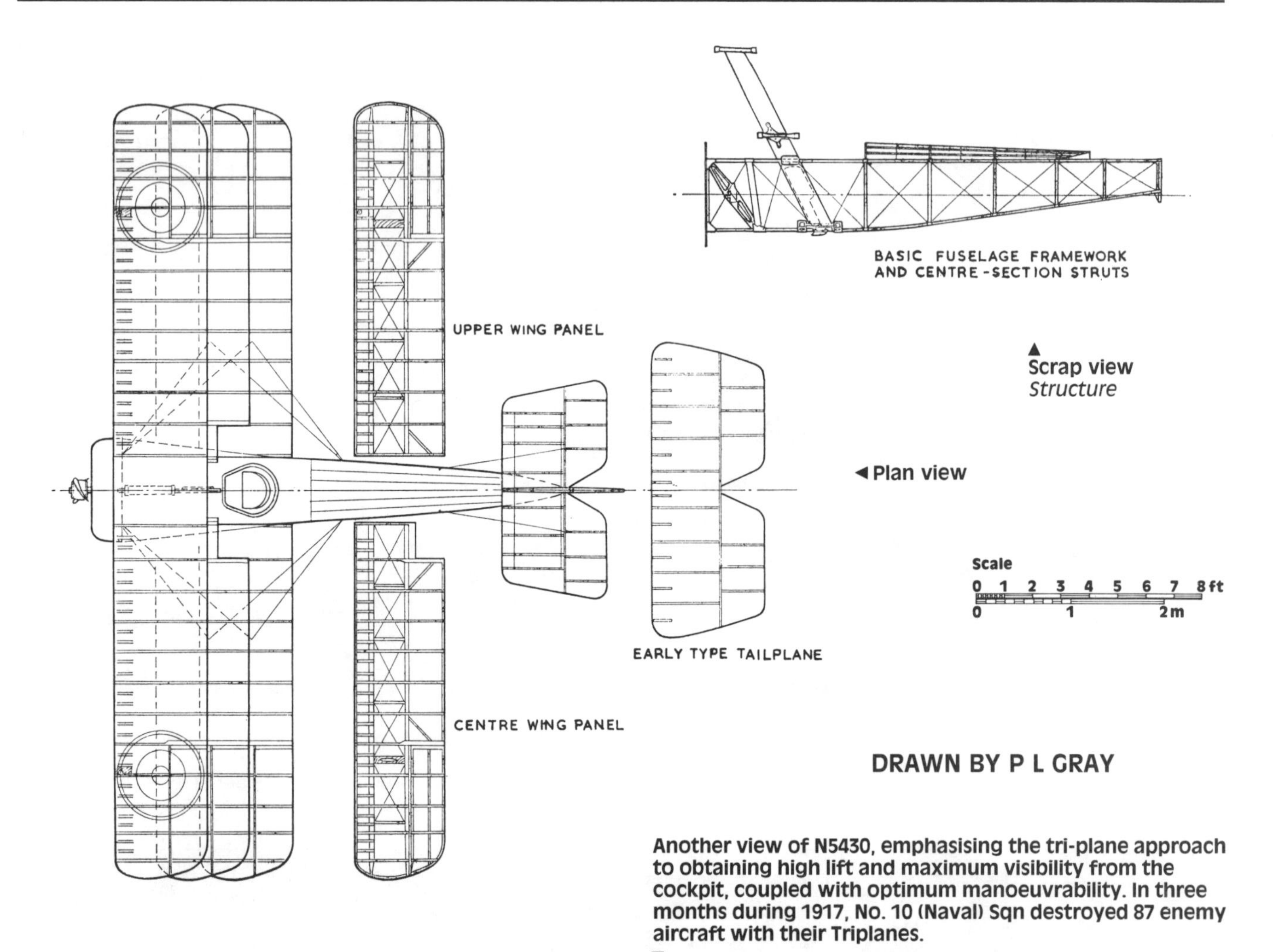

▲ Scrap view
Structure

◄ Plan view

DRAWN BY P L GRAY

Another view of N5430, emphasising the tri-plane approach to obtaining high lift and maximum visibility from the cockpit, coupled with optimum manoeuvrability. In three months during 1917, No. 10 (Naval) Sqn destroyed 87 enemy aircraft with their Triplanes.
▼

Sopwith 1½ Strutter

Country of origin: Great Britain.
Type: Two-seat fighter.
Powerplant: One Clerget rotary engine rated at 110hp (later 130hp).
Dimensions: Wing span 33ft 6in *10.21m*; length 25ft 3in *7.70m*; height 10ft 3in *3.12m*.
Weights: Empty 1259lb *571kg*, (130hp engine) 1305lb *592kg*; loaded 2150lb *975kg*.
Performance: Maximum speed 100mph *161kph* at 6500ft *1980m*; service ceiling 15,500ft *4720m*.
Armament: One fixed Vickers machine gun and one flexibly mounted Lewis machine gun.
Service: First flight December 1915; service entry early 1916.

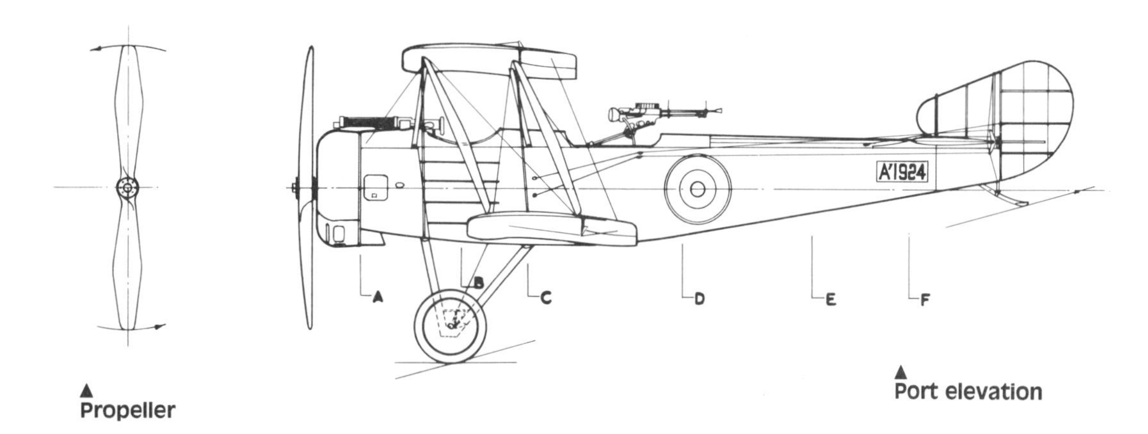

▲ Propeller

▲ Port elevation

An unarmed 1½ Strutter with polished metal cowling and translucent dope on the wings (showing some light through, although not clearly as the centre panels in some aircraft – see other photos). (P L Gray) ▶

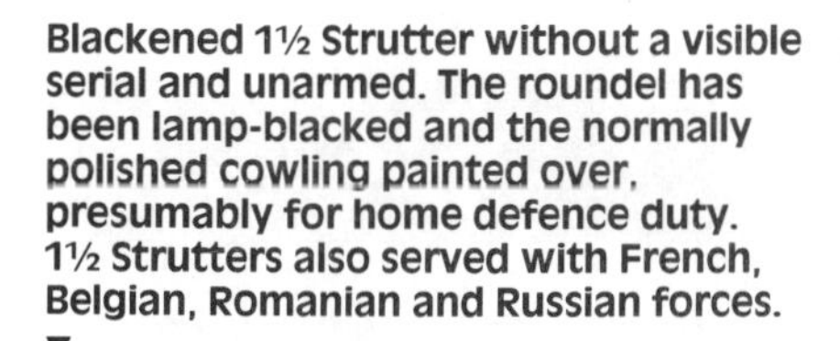

Blackened 1½ Strutter without a visible serial and unarmed. The roundel has been lamp-blacked and the normally polished cowling painted over, presumably for home defence duty. 1½ Strutters also served with French, Belgian, Romanian and Russian forces. ▼

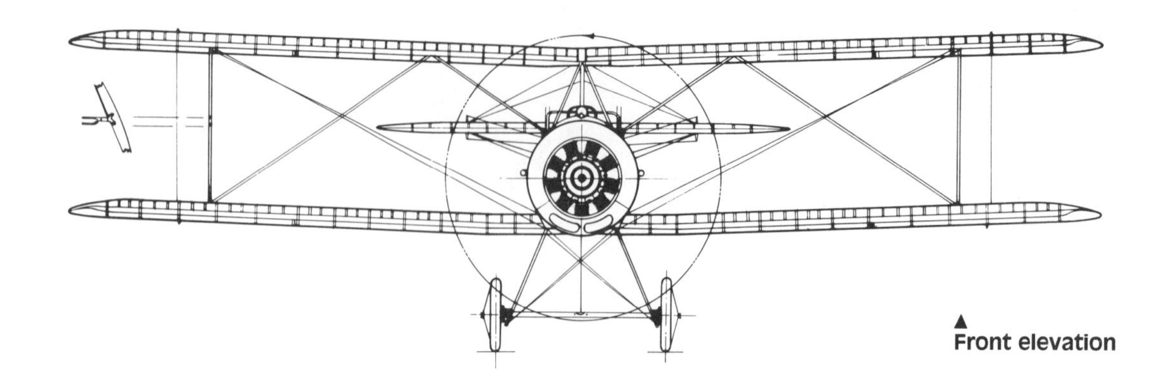

▲
Front elevation

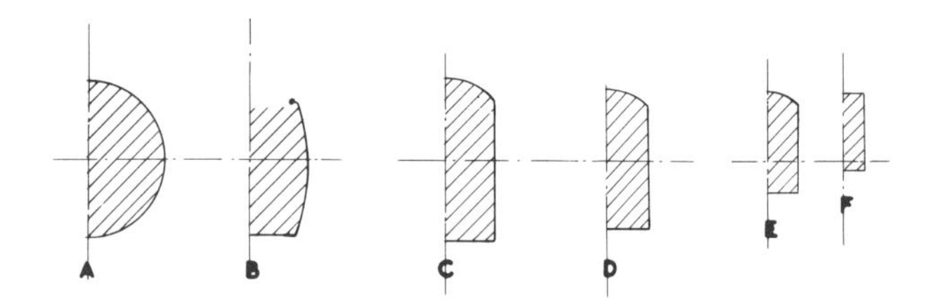

◀Fuselage cross-sections

▲
Captured by the Germans in 1916, this 1½ Strutter shows the clear panels of the upper centre section and the single forward-firing gun. (P L Gray)

Unarmed 9378, crashed, with oil-soaked covering on the forward cockpit sides and air brakes firmly in the 'up' position. (P L Gray)
▼

Wing cross-section ▼

Plan view ▼

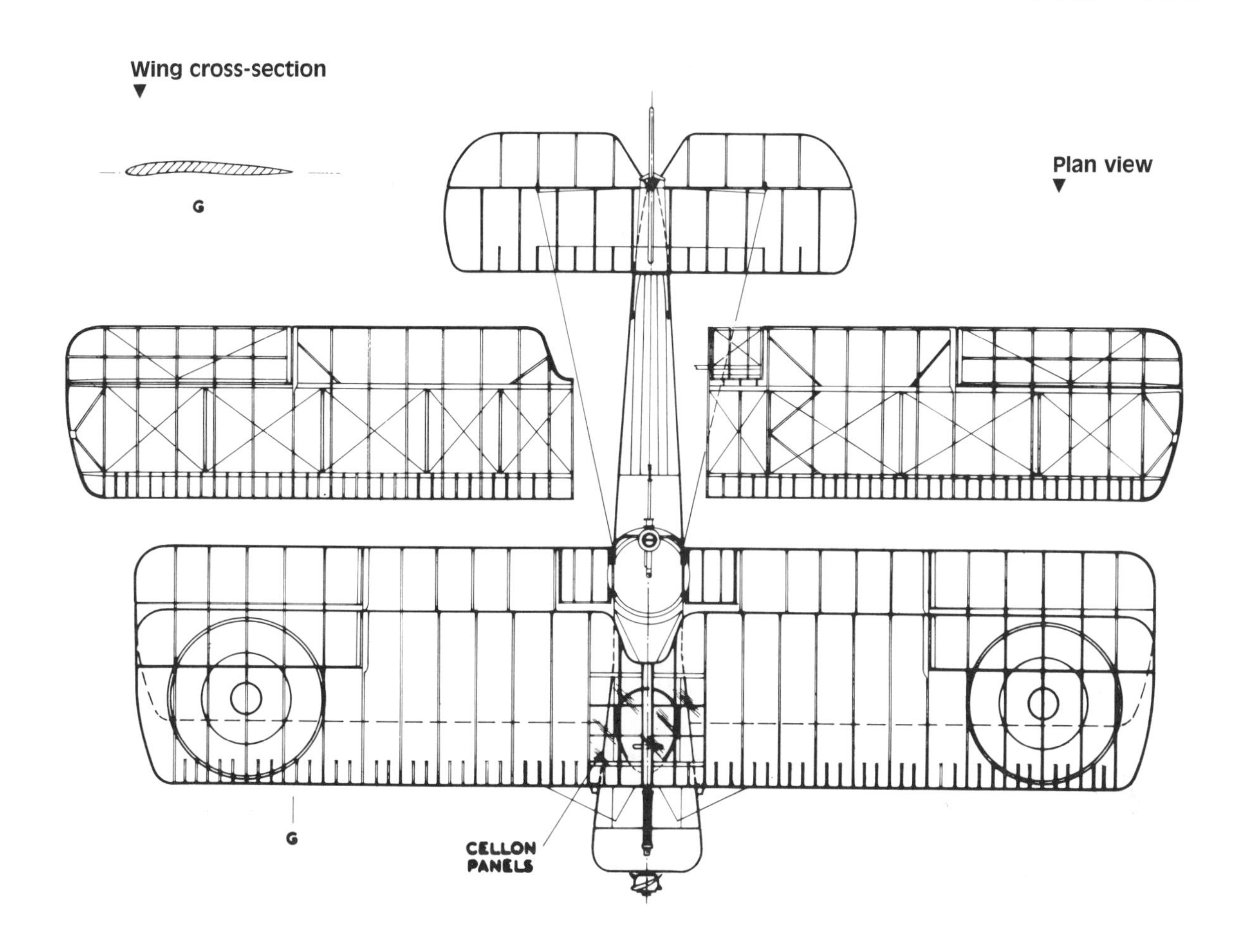

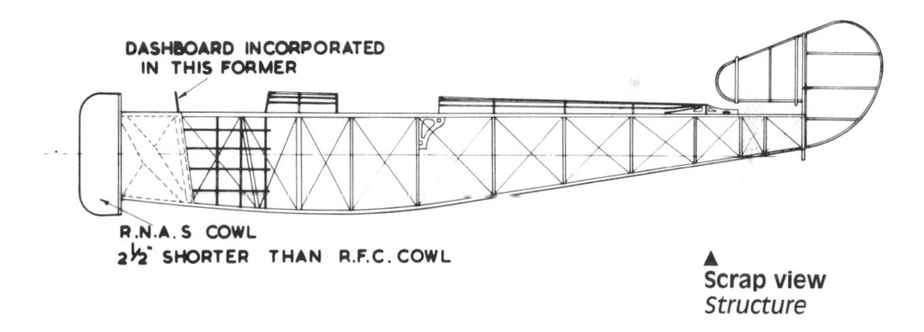

▲
Scrap view
Structure

Colour notes
Initially 1½ Strutters were finished in overall clear doped fabric but during late summer 1916 were appearing with upper and side surfaces finished in khaki-green. Cowlings and metal nose panels were sometimes bright, natural finish but were more usually painted grey, as were the metal undercarriage and centre-section struts. Wheel discs were either natural linen or khaki-green. Roundels (blue outermost) were applied to all four wing tips, those on the upper surface being narrowly outlined in white; fuselage roundels were not always outlined.

DRAWN BY P L GRAY

Experimental 1½ Strutter with skid chassis, hydrovane and flotation bags. Note the extra strut on A5952. (J M Bruce/G S Leslie) ►

Breguet XIV A2 and B2

Country of origin: France.
Type: Two-seat reconnaissance aircraft and (B2) day bomber.
Powerplant: One Renault 12F engine rated at 300hp, Renault rated at 310hp or Fiat rated at 285hp.
Dimensions: Wing span 47ft 3in *14.40m*; length 29ft 7in *9.02m*; height 10ft 8½in *3.26m*.
Weights: Empty 2222lb *1008kg*, (B2) 2730lb *1238kg*; loaded 3380lb *1533kg*, (B2) 4300lb *1950kg*.
Performance: Maximum speed 118mph *190kph* at sea level, (B2) 115mph *185kph* at sea level; time to 16,400ft *5000m*, 22min, (B2) 47min; service ceiling 20,000ft *6100m*, (B2) 18,900ft *5760m*; endurance 4.5hr.
Armament: Bomb load up to about 660lb *300kg* (B2 only), plus one fixed Vickers machine gun and single or twin flexibly mounted Lewis or Hotchkiss machine guns.
Service: First flight November 1916; service entry 1917.

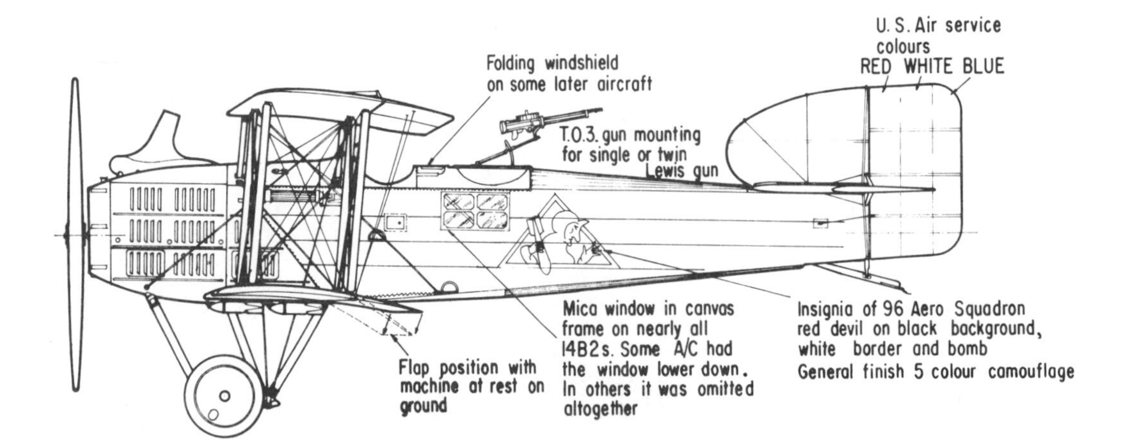

Port elevation, XIV B2

A 1917 Breguet XIV (300hp Renault) displays its enormous ailerons on upper wings and depressed 'flaps' on lower. (M Bayet)

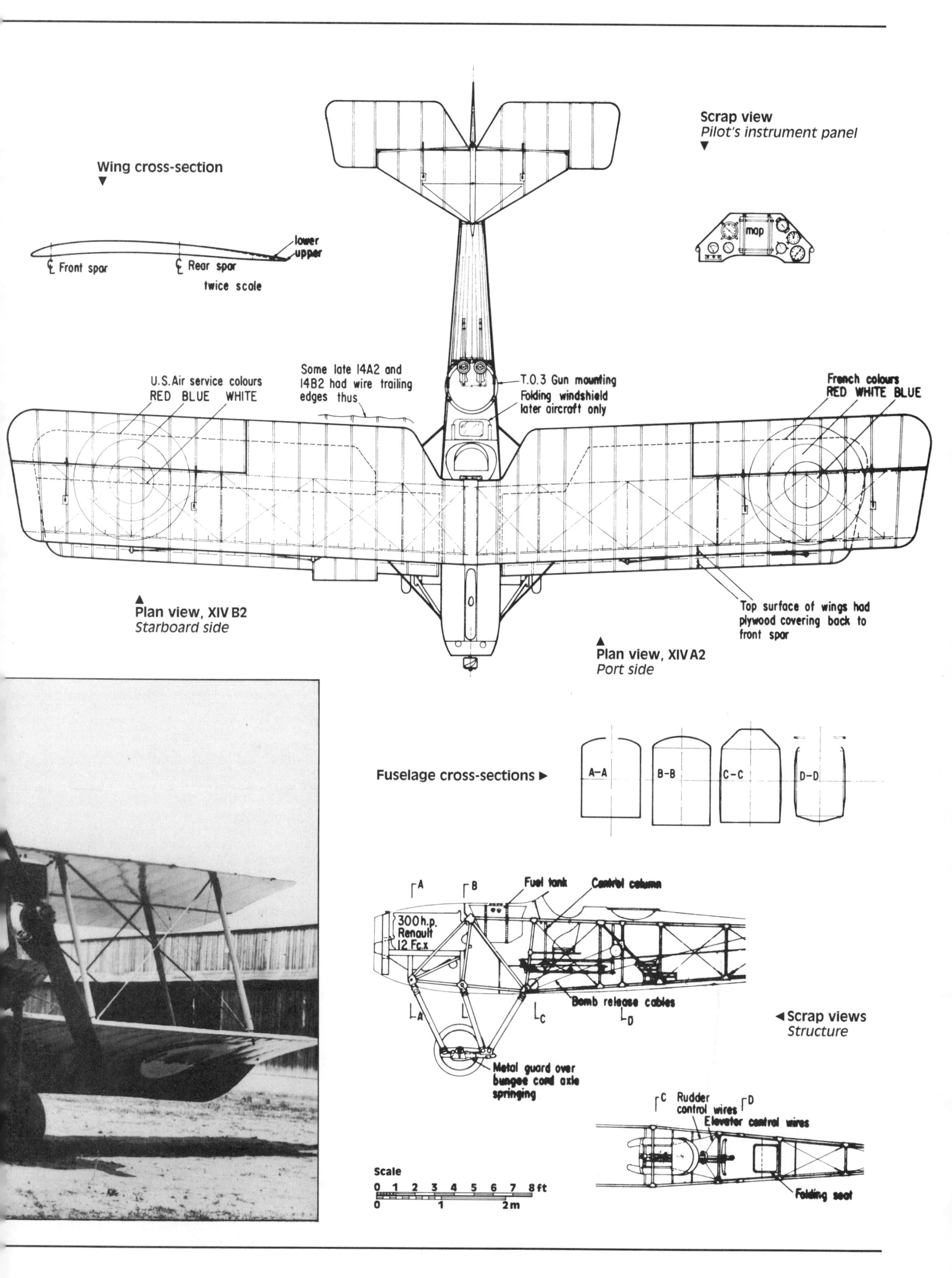
Wing cross-section
lower
upper
℄ Front spar
℄ Rear spar
twice scale
Scrap view
Pilot's instrument panel
map
U.S.Air service colours
RED BLUE WHITE
Some late 14A2 and
14B2 had wire trailing
edges thus
T.O.3 Gun mounting
Folding windshield
later aircraft only
French colours
RED WHITE BLUE
Top surface of wings had
plywood covering back to
front spar
Plan view, XIV B2
Starboard side
Plan view, XIV A2
Port side
Fuselage cross-sections
A-A
B-B
C-C
D-D
Fuel tank
Control column
300 h.p.
Renault
12 Fc.x
Bomb release cables
Metal guard over
bungee cord axle
springing
Scrap views
Structure
Rudder
control wires
Elevator control wires
Folding seat
Scale
0 1 2 3 4 5 6 7 8 ft
0 1 2 m

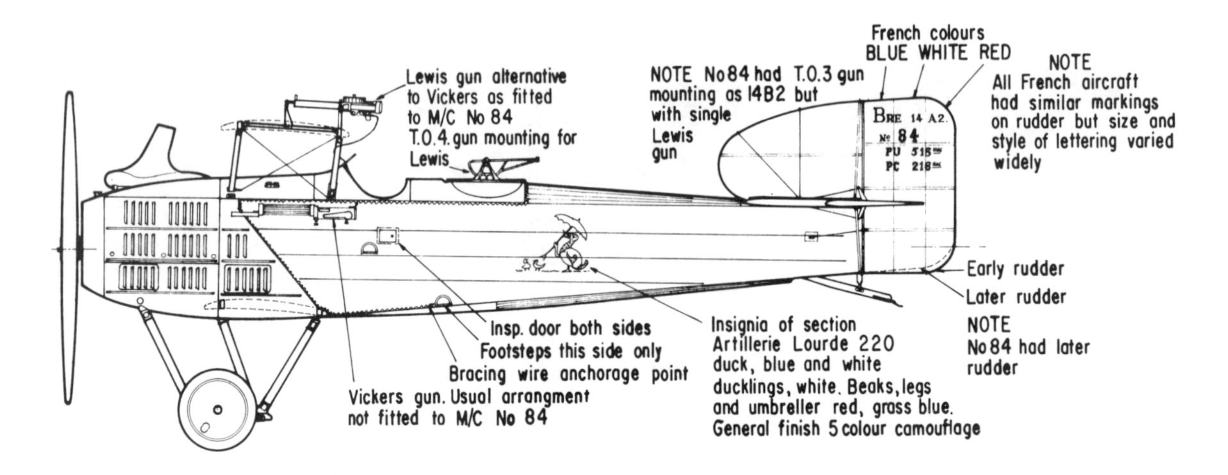

▲
Port elevation, XIV A2

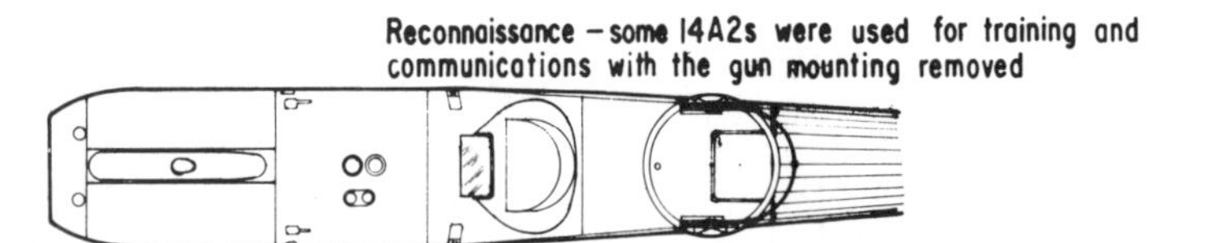

◄**Scrap plan view**
Forward fuselage

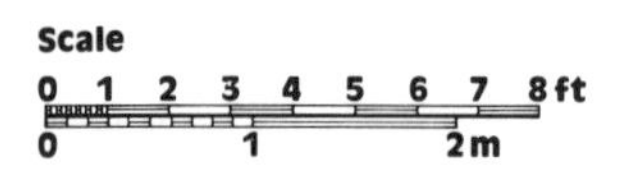

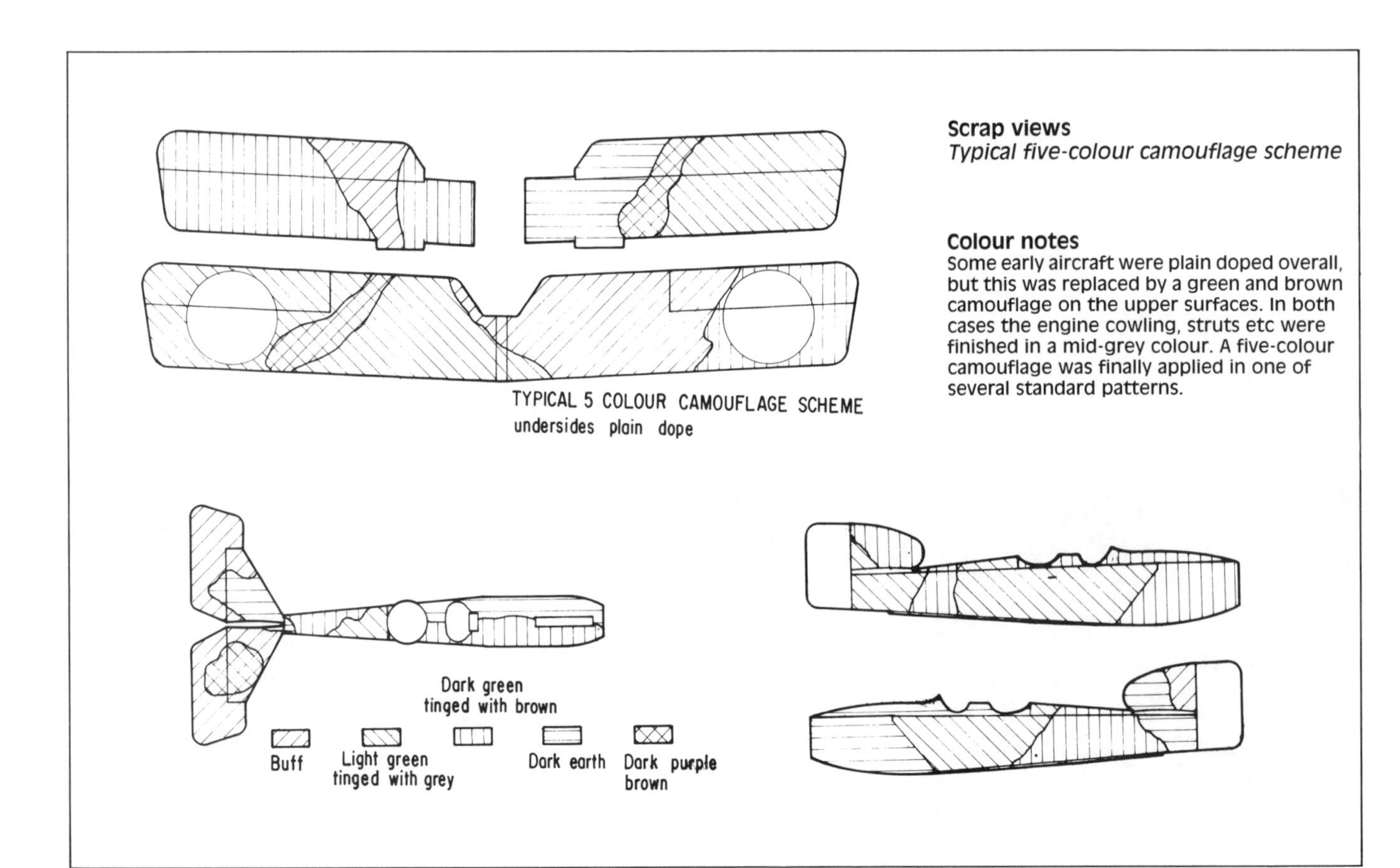

Scrap views
Typical five-colour camouflage scheme

Colour notes
Some early aircraft were plain doped overall, but this was replaced by a green and brown camouflage on the upper surfaces. In both cases the engine cowling, struts etc were finished in a mid-grey colour. A five-colour camouflage was finally applied in one of several standard patterns.

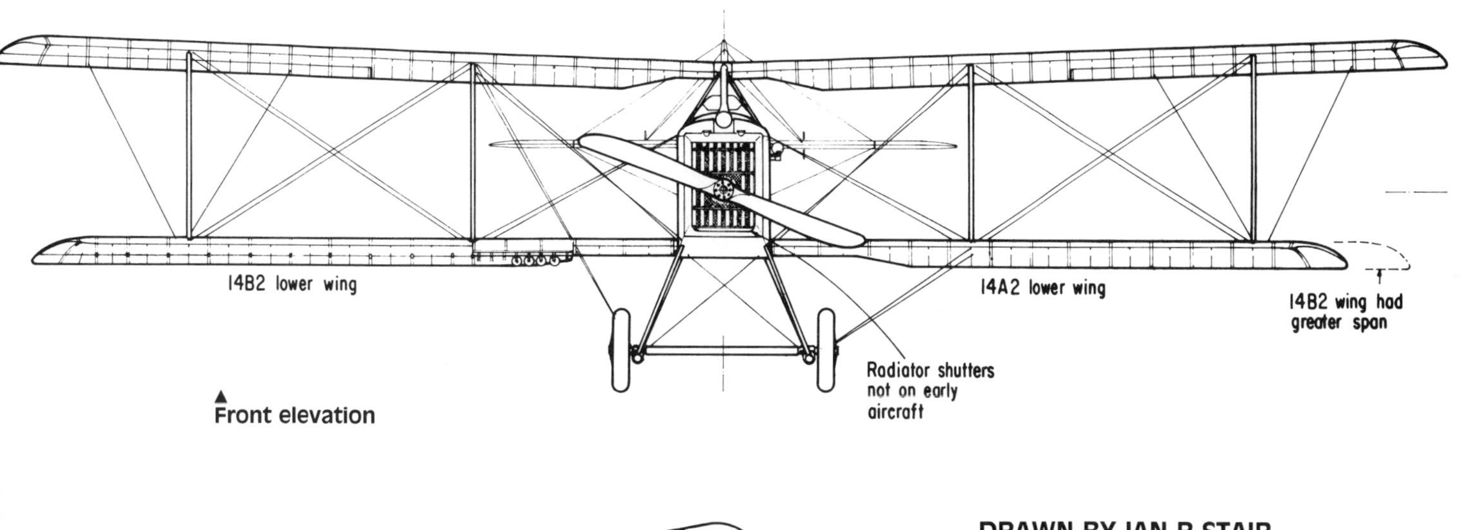

▲ Front elevation

DRAWN BY IAN R STAIR

Underplan, XIV A2 ►
Port side

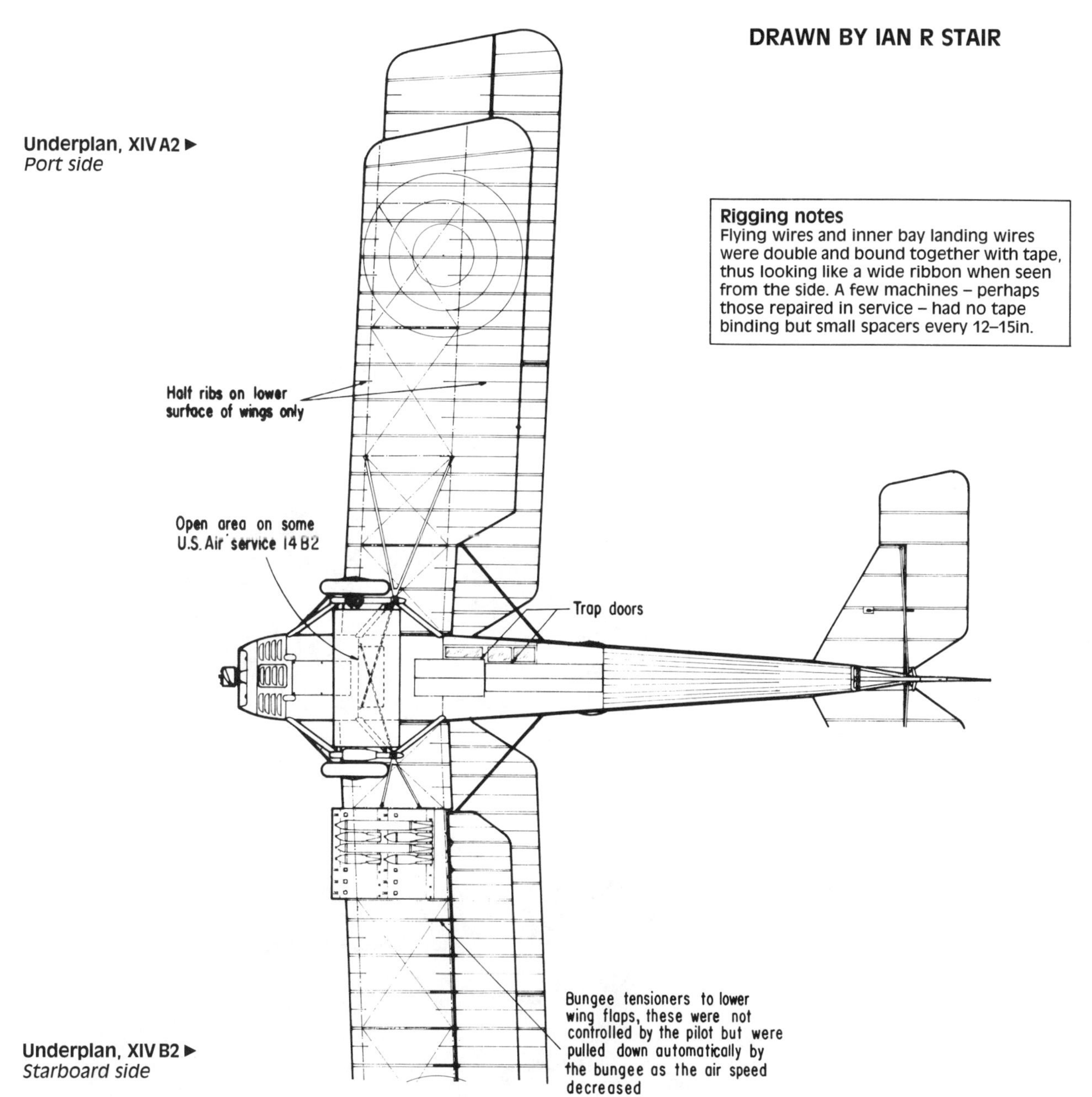

Rigging notes
Flying wires and inner bay landing wires were double and bound together with tape, thus looking like a wide ribbon when seen from the side. A few machines – perhaps those repaired in service – had no tape binding but small spacers every 12–15in.

Underplan, XIV B2 ►
Starboard side

Morane Saulnier 35EP

Country of origin: France.
Type: Two-seat primary trainer.
Powerplant: One Le Rhône rotary engine rated at 80hp.
Dimensions: Wing span 34ft 7¾in *10.56m*; length 22ft 1¾in *6.75m*; height 11ft 9¾in *3.60m*; wing area 193.75 sq ft *18.0m²*.
Weights: Empty 1014lb *460kg*; loaded 1543lb *700kg*.
Performance: Maximum speed 84.25mph *135.5kph*; time to 3280ft *1000m*, 6.55min.
Armament: None.
Service: Service entry 1922.

Port elevation ►

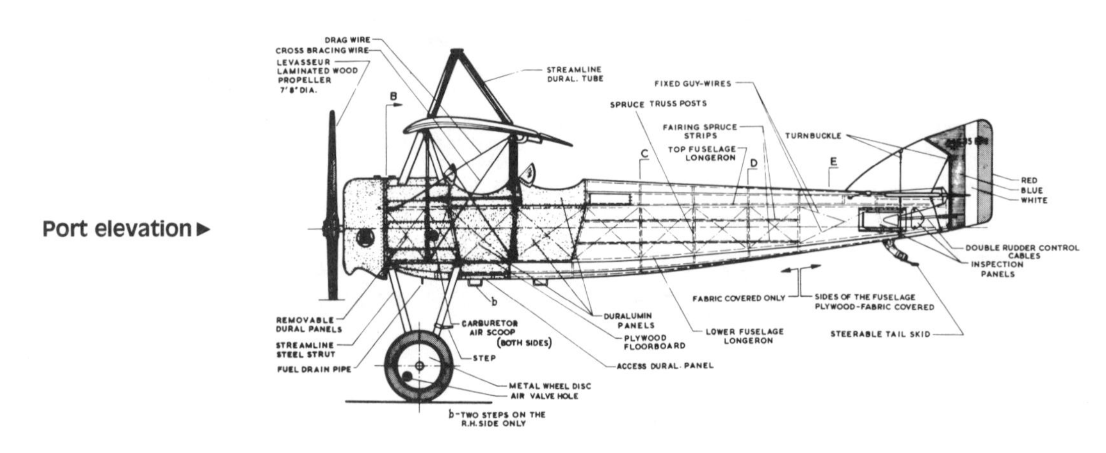

Scale
0 1 2 3 4 5 6 7 8ft
0 1 2m

TRADE MARK "MS" WAS CUT OUT OF DURAL. SHEET AND PLACED ON ENGINE COWLING, ON BOTH SIDES. THE COLOUR WAS RED

All-green fabric and polished metal panels on the Morane 35 *'Ecole Premiere'* coupled with large Polish insignia make this wartime-designed (but postwar-utilised) trainer an attractive modelling subject. (T Zychiewicz)
▼

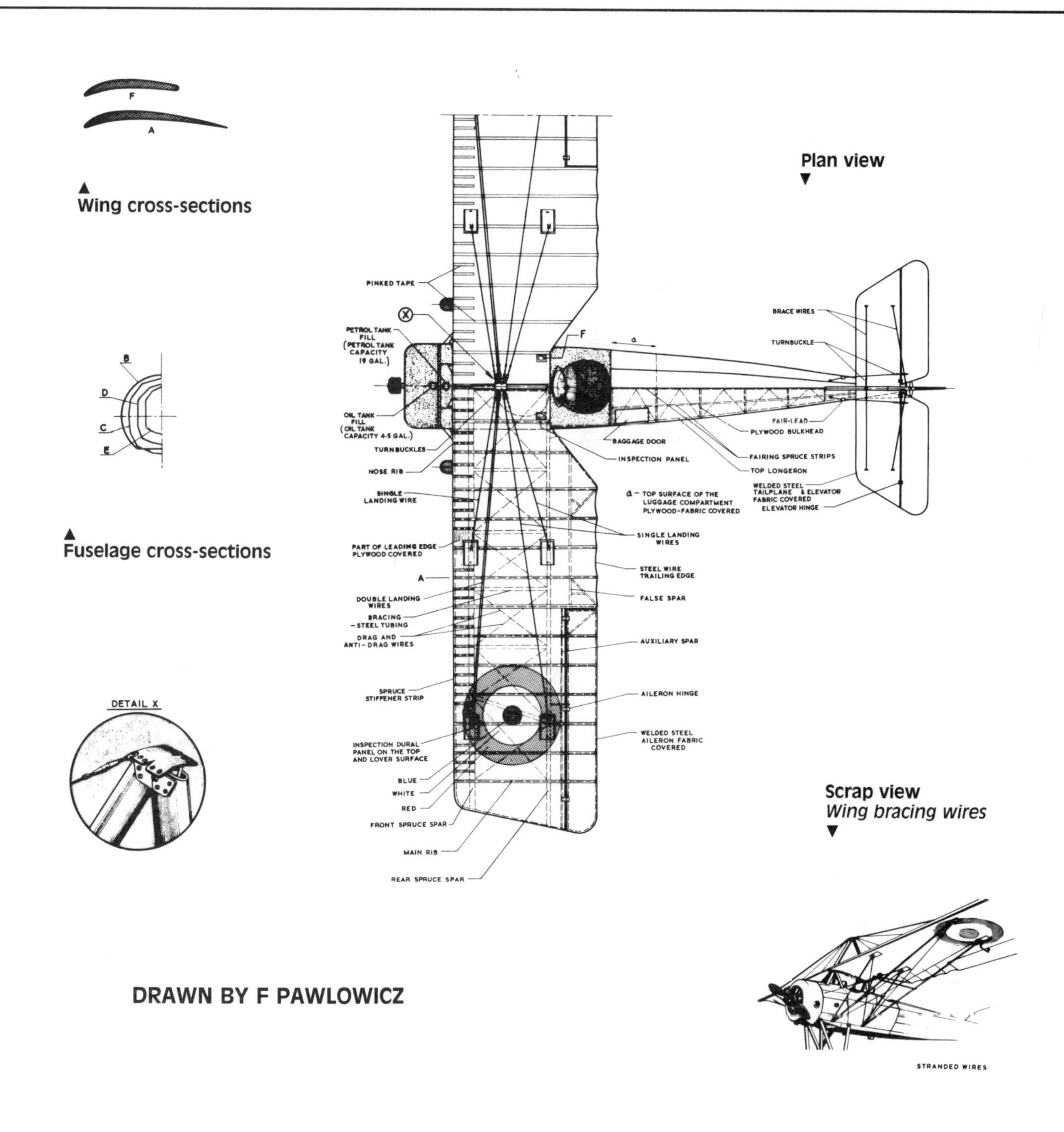

▲
Wing cross-sections

▲
Fuselage cross-sections

Plan view
▼

Scrap view
Wing bracing wires
▼

DRAWN BY F PAWLOWICZ

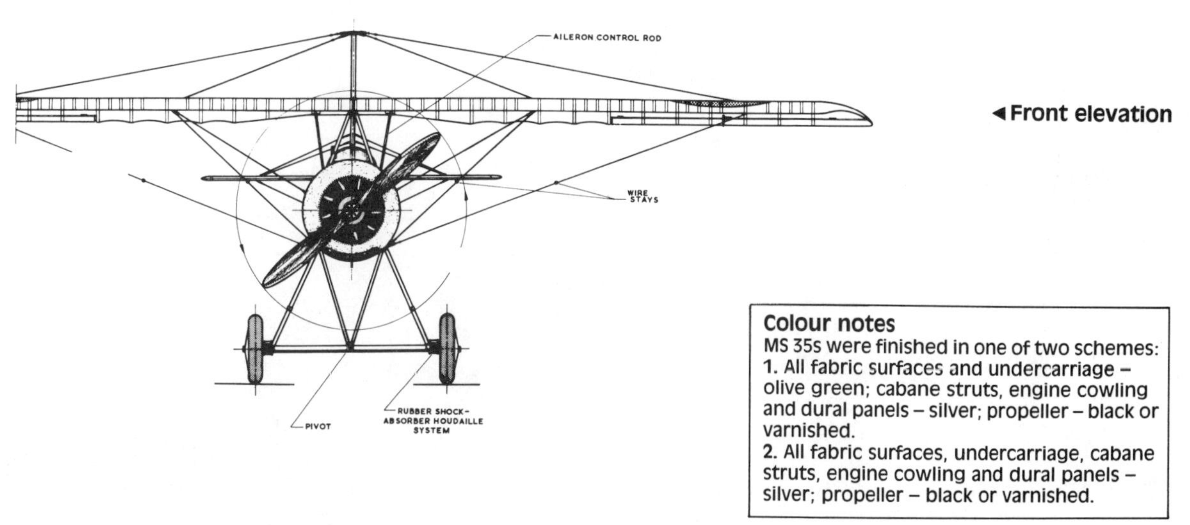

◄Front elevation

Colour notes
MS 35s were finished in one of two schemes:
1. All fabric surfaces and undercarriage – olive green; cabane struts, engine cowling and dural panels – silver; propeller – black or varnished.
2. All fabric surfaces, undercarriage, cabane struts, engine cowling and dural panels – silver; propeller – black or varnished.

Liaison machine for the 6th Regiment at Krakow in 1925, this EP and adjacent crew illustrate the generous proportions for a modest 80hp Le Rhône rotary engine. Morane parasol designs were remarkably efficient and went on to extended development into the 1930s. (F Pawlowicz)►

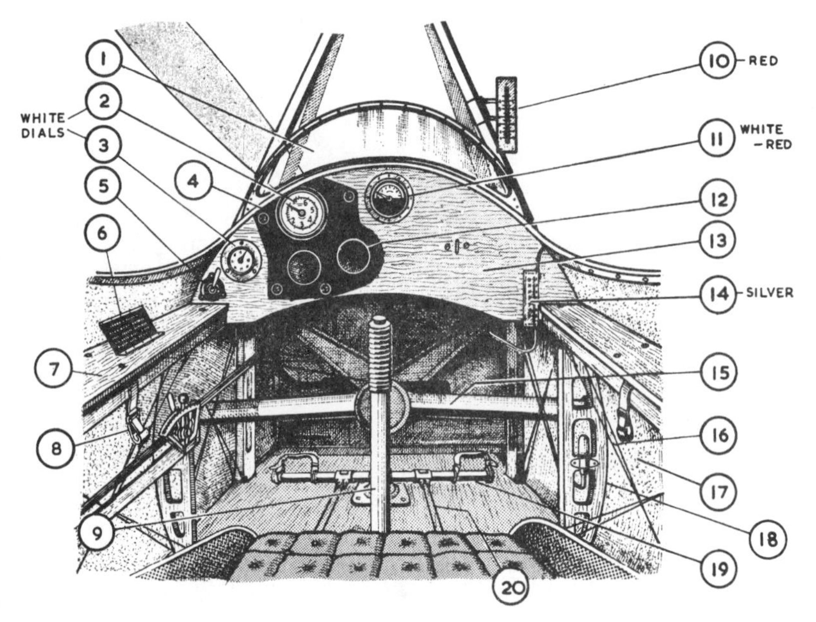

Key to sketches

Front cockpit: 1. Plexi windshield. **2.** Altimeter. **3.** Tachometer. **4.** Ebony panel (black). **5.** Ignition switch. **6.** Instruction plate (black). **7.** Wooden bearing. **8.** Safety strap holder. **9.** Control stick. **10.** Incidence indicator. **11.** Fuel gauge. **12.** Two holes for optical instruments. **13.** Plywood instrument panel. **14.** Oil indicator. **15.** Carburettor air pipes. **16.** Guy wires. **17.** Dural panel. **18.** Wooden former. **19.** Rudder bar. **20.** Rudder bar connection cables.

Rear cockpit: 1. Opening through which instrument panel in front cockpit could be seen during solo flight. **2.** Contact breaker. **3.** Ignition switch. **4.** Throttle. **5.** Mixture control. **6.** Front seat. **7.** Spruce truss. **8.** Push button switch. **9.** Steel tubing of structure. **10.** Rudder control cables. **11.** Aileron control rod.

Note: All elements of fuselage wooden structure (formers, floor, seats, instrument panels etc) were varnished.

Sketch section

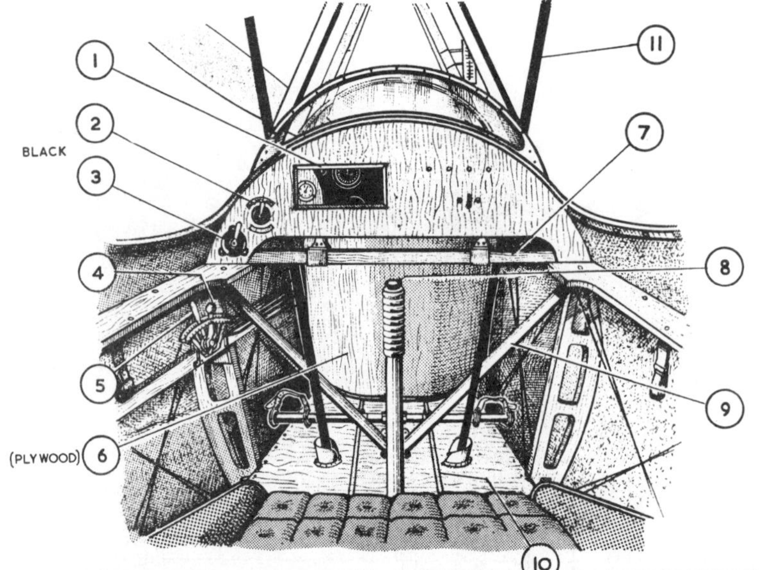

Nieuport 17C

Country of origin: France.
Type: Single-seat scout.
Powerplant: One Le Rhône rotary engine rated at 110hp.
Dimensions: Wing span 27ft 0in *8.23m*; length 19ft 6in *5.94m*; height 7ft 0in *2.13m*; wing area 158.8 sq ft *14.75m²*.
Weights: Empty 825lb *374kg*; loaded 1233lb *559kg*.
Performance: Maximum speed 107mph *172kph* at 6500ft *1980m*; time to 6500ft, 5.5min; service ceiling 17,400ft *5300m*; endurance 2hr.
Armament: One fixed Lewis or (later aircraft) Vickers machine gun, plus four Le Prieur rockets.
Service: First flight 1915; service entry 1916.

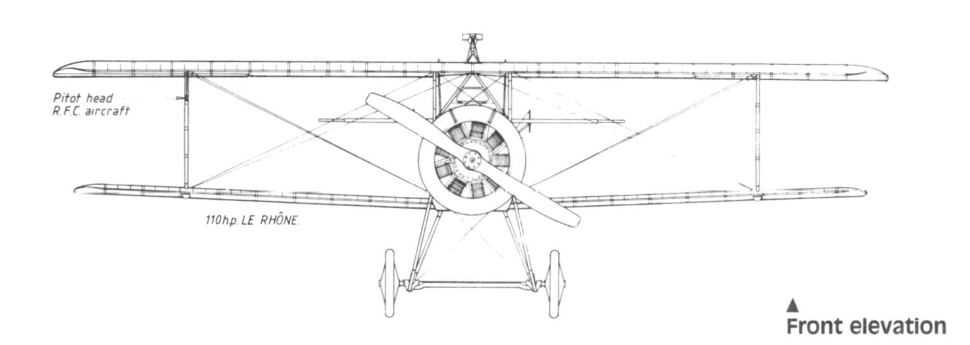

▲ **Front elevation**

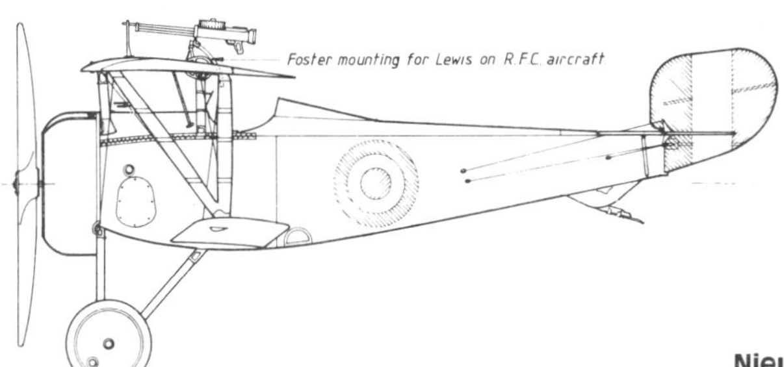

◄ **Port elevation**
RFC aircraft with Lewis gun

Nieuport 17 of the French *Aviation Militaire*, showing the standard armament for the type favoured by that service – the single fixed, synchronised Vickers machine gun mounted on the centreline of the forward fuselage. (A Imrie)
▼

▲
Nieuport 17 captured intact by the Germans in September 1916 and extensively flown at the Adlershof test centre. The fuselage decoration and the single Vickers armament indicate a French-operated machine, and note that the completely circular engine cowling has joining ribs at the top and not on the sides as shown in the other photograph. (A Imrie)

Stork insignia on the fuselage side identifies this Nieuport 17CI as belonging to *Escadrille N3 'Les Cigognes'*, in 1916. Other than rudder stripes, no national markings are carried, and wings and tail are outlined in undersurface blue. (H Woodman)
▼

Starboard elevation
French aircraft with Vickers gun
▼

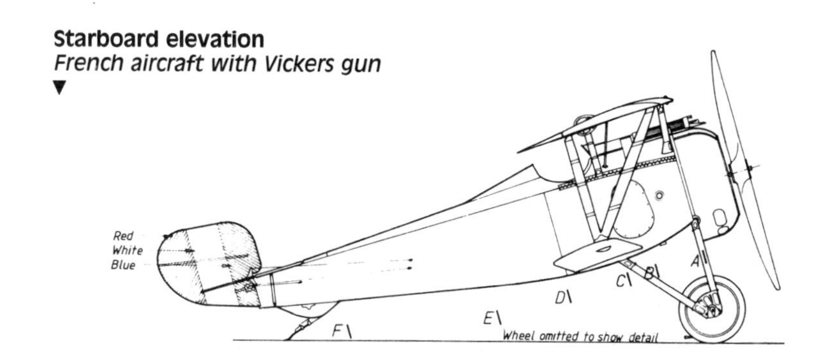

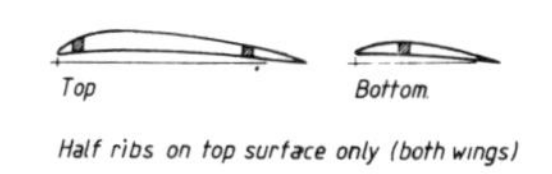

▲
Wing cross-sections
Half-ribs on top surface only (both wings)

Colour notes
Most Nieuport 17s were overall silver-grey but some early aircraft had a dark camouflage finish on the upper surfaces and were clear doped underneath.

Fuselage cross-sections ►

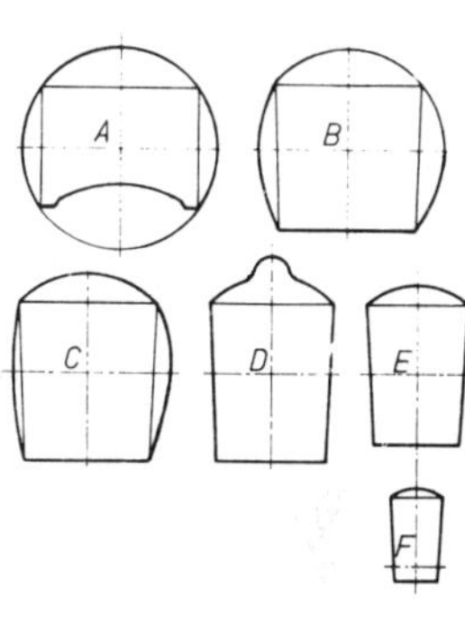

Plan view
▼

Scrap plan view
Lower wing
▼

DRAWN BY IAN R STAIR

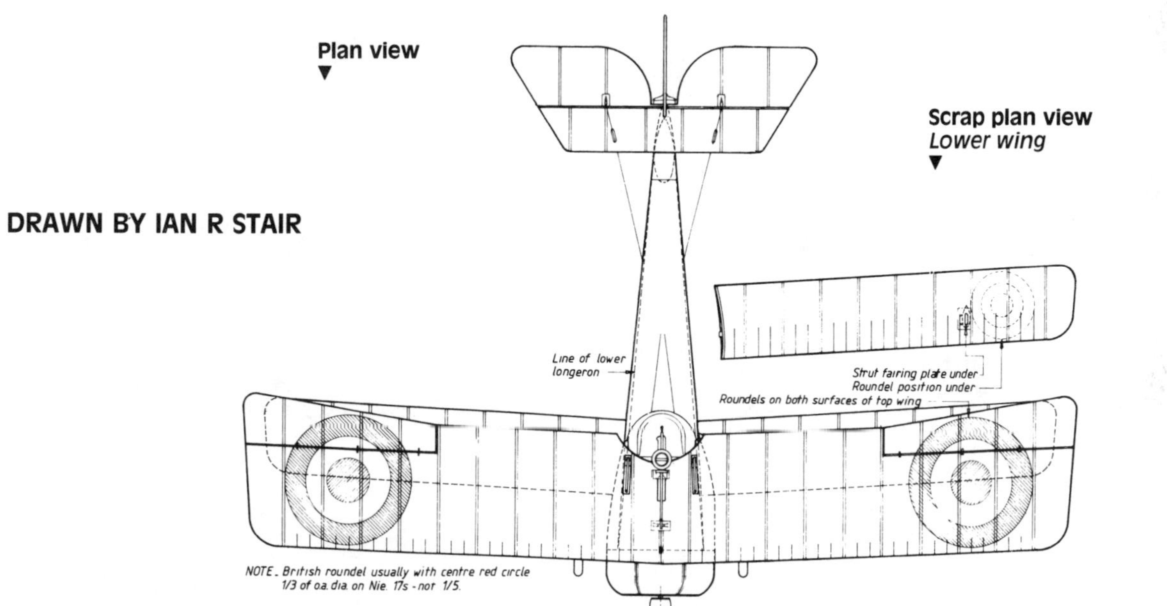

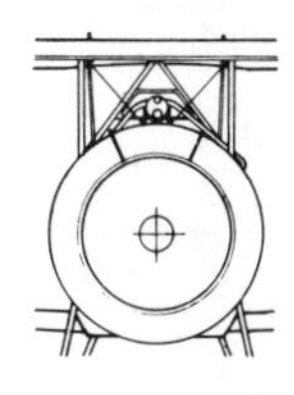

Scale

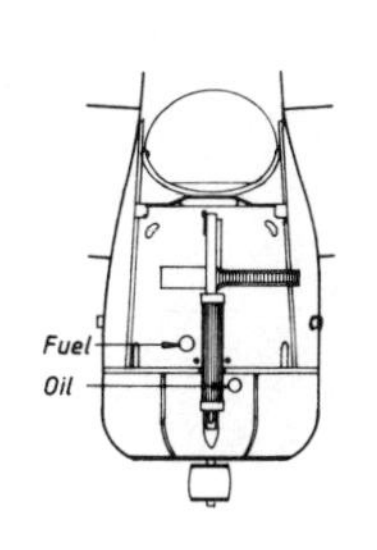

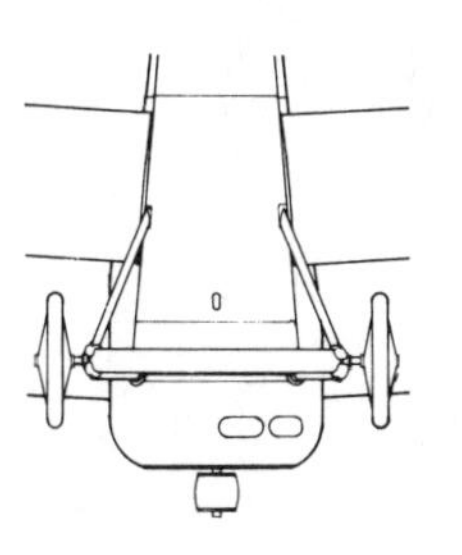

Scrap views ▲ ►
Vickers gun installation

▲
Scrap underplan

▲
Type 17c of 1916 with the Gnome 80hp rotary, bearing the famous markings of *Lt* Charles Nungesser. Red, white and blue diagonals on upper wing were to warn off Allied aircraft after he was attacked and obliged to retaliate with fatal results. (Musée de l'Air)

Belgian ace Edmund Thieffrey in the cockpit of his Nieuport 17 bearing the comet markings of No 1 Sqn, Belgian Air Service, 1917. (H Woodman) ▶

Nieuport 17s also served with the Russian forces: this example from the Red Air Fleet would have been from the 1919–20 period. (H Woodman)
▼

Nieuport 28

Country of origin: France.
Type: Single-seat scout.
Powerplant: One Gnome Monosoupape 9N rotary engine rated at 150hp.
Dimensions: Wing span 26ft 9¼in *8.16m*; length 21ft 0in *6.40m*; height 8ft 1in *2.46m*.
Weights: Empty 961lb *436kg*; loaded 1539lb *698kg*.
Performance: Maximum speed 128mph *206kph*; service ceiling 19,700ft *6000m*; endurance about 2hr.
Armament: Two fixed Vickers machine guns.
Service: Service entry 1917.

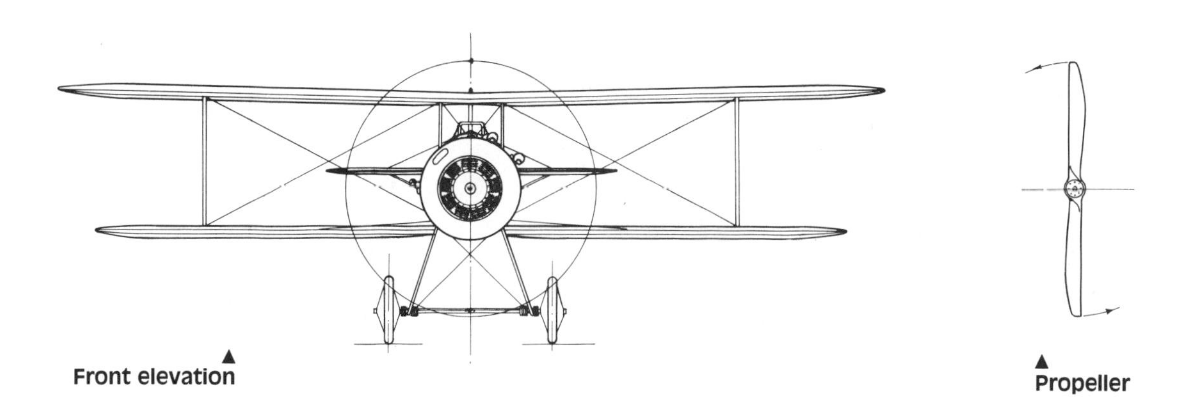

▲ Front elevation

▲ Propeller

Colour notes
Upper and side surfaces camouflaged in irregular patches of light grey-green, dark olive green, light earth and dark brown. Colours on fuselage were in order: dark brown (nose), light green, olive green and a small patch of dark brown under tailplane. Undersurfaces of wings and tailplane were natural fabric, clear-doped and varnished. Interplane struts were natural varnished spruce, while undercarriage and centre-section struts were painted green. Roundels were carried on both surfaces of upper wings and underneath lower wings but not on fuselage sides.

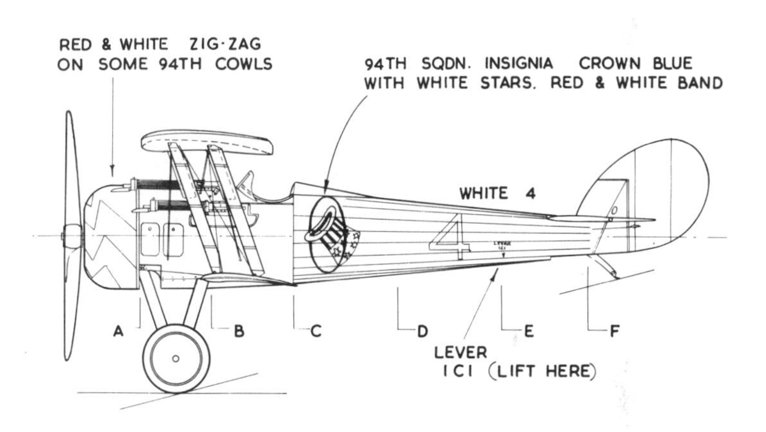

▲ Port elevation

Fuselage cross-sections ▼

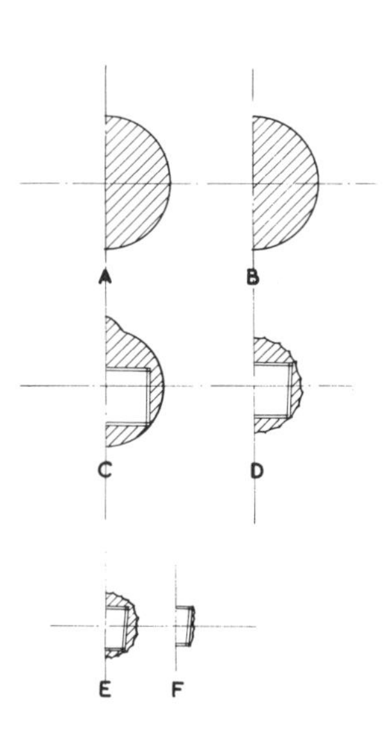

Capt Reed Chambers in his Nieuport 28 at Toul in 1918, well protected in the cockpit and adequately armed. (E van Gorder) ▼

▲
Similar, but not quite the same! This rebuilt N28, in the Paul Mantz collection, is a veteran of several films and is still preserved elsewhere. It has had drag struts added to the interplane bracing and has received replacement wheels. (R Linn)

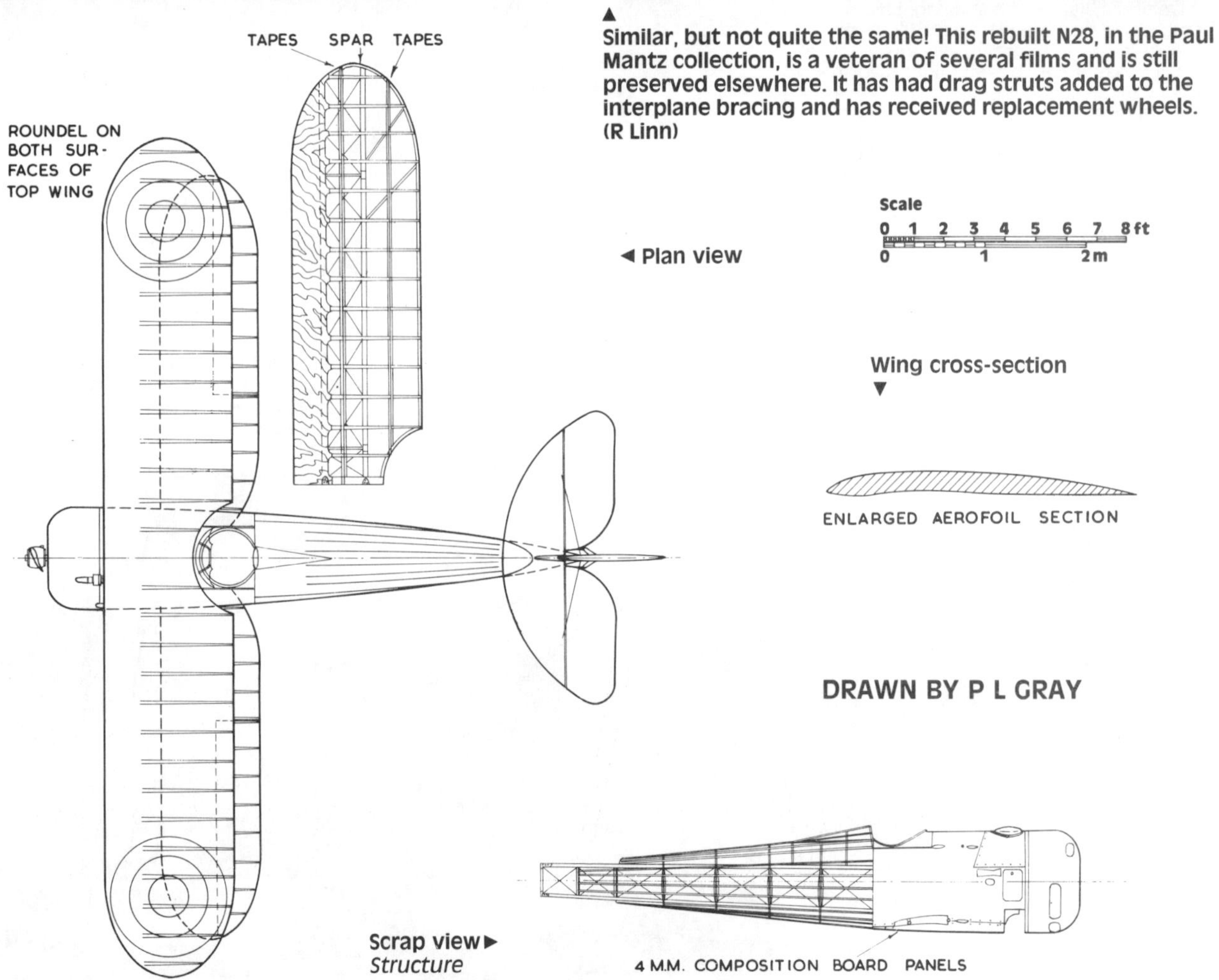

SPAD XIII

Country of origin: France.
Type: Single-seat scout.
Powerplant: One Hispano-Suiza 8Ba engine rated at 220hp or (later aircraft) 8BEc rated at 235hp.
Dimensions: Wing span 26ft 4in *7.90m*; length 20ft 8in *6.30m*; height 7ft 6in *2.29m*; wing area 215 sq ft *20.0m²*.
Weights: Loaded 1807lb *820kg*.
Performance: Maximum speed 134.4mph *216.4kph* at sea level, (8BEc) 138.5mph *223kph* at sea level; time to 6500ft *1980m*, 5.25min, (8BEc) 4.5min; service ceiling 22,300ft *6800m*, (8BEc) 22,350ft *6815m*; endurance 1.66hr.
Armament: Two fixed Vickers machine guns, plus two 25lb *11.3kg* bombs.
Service: First flight summer 1917; service entry August 1917.

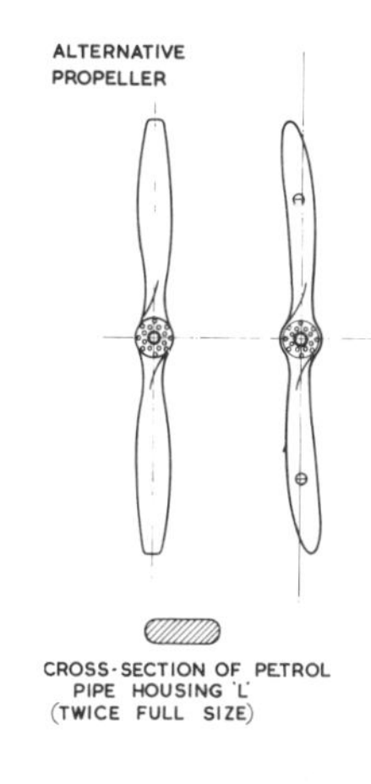

▲
Propeller details

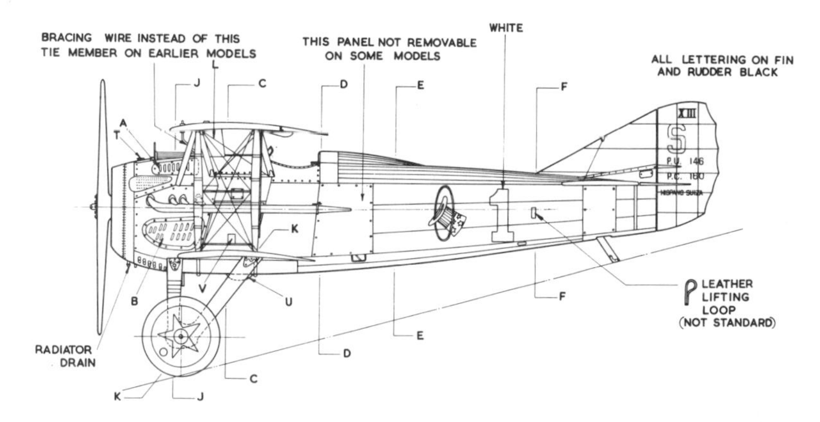

DRAWN BY G A G COX

▲
Port elevation

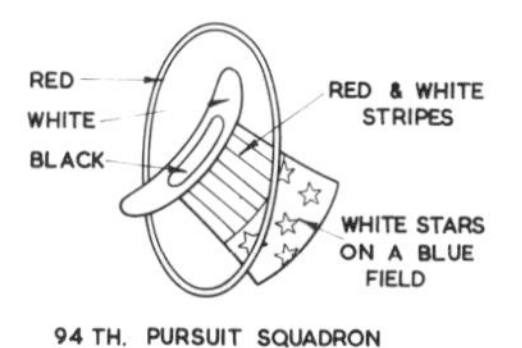

Scrap view ▶
Fuselage markings

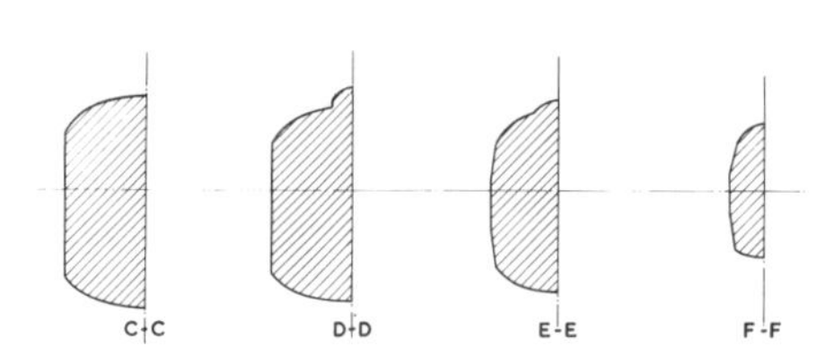

▲
Fuselage cross-sections

The SPAD XIIIC-1 was selected by the Bolling Commission to be manufactured by Curtiss; 2000 were to be produced, but these were cancelled and 893 were procured from France. With the 220hp Hispano-Suiza engine, the XIII was a formidable (and colourful) fighter with the AEF. It continued in US service for some years after the war. (USAF)
▼

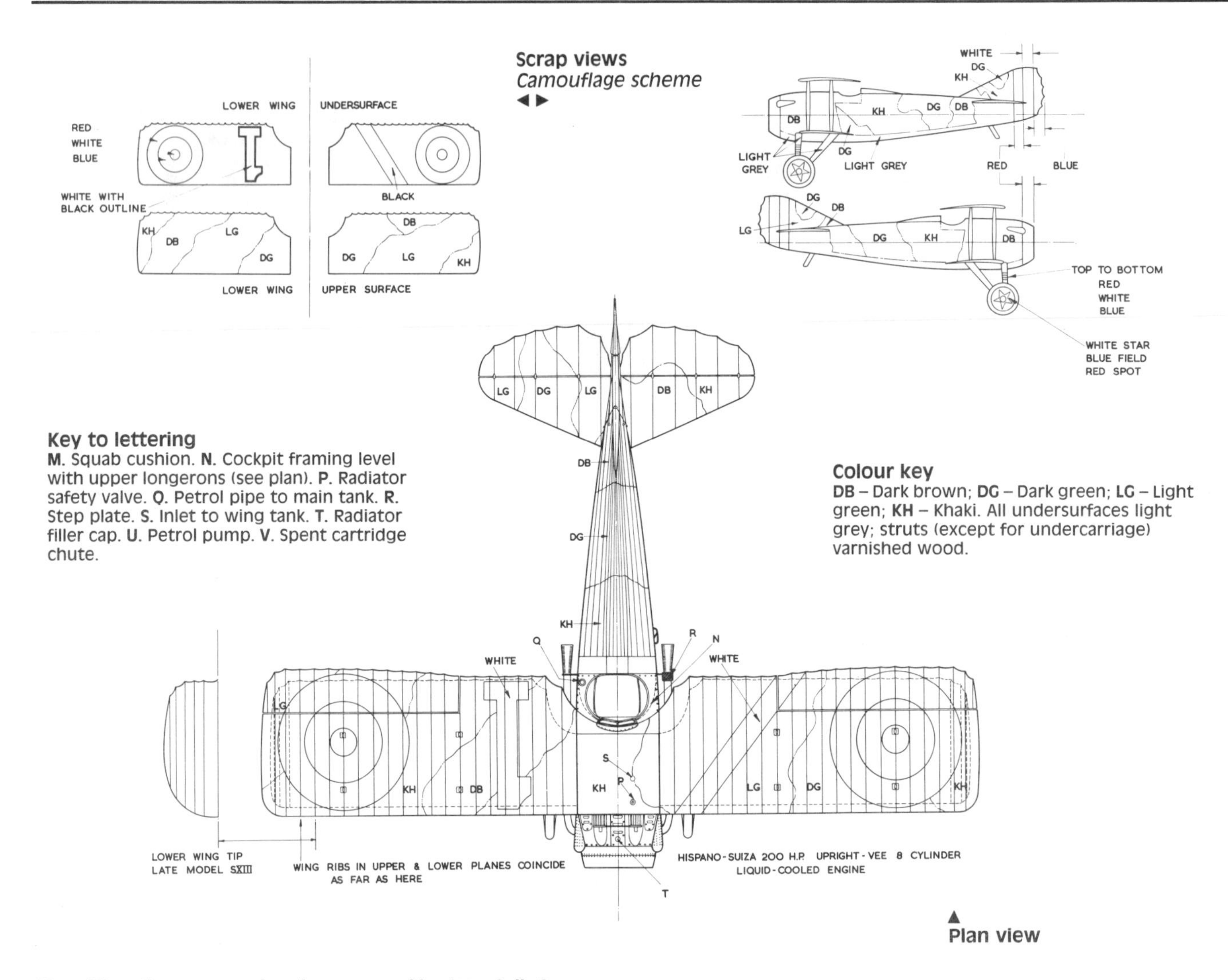

Scrap views
Camouflage scheme
◀ ▶

Key to lettering
M. Squab cushion. **N.** Cockpit framing level with upper longerons (see plan). **P.** Radiator safety valve. **Q.** Petrol pipe to main tank. **R.** Step plate. **S.** Inlet to wing tank. **T.** Radiator filler cap. **U.** Petrol pump. **V.** Spent cartridge chute.

Colour key
DB – Dark brown; **DG** – Dark green; **LG** – Light green; **KH** – Khaki. All undersurfaces light grey; struts (except for undercarriage) varnished wood.

▲
Plan view

Magnificently preserved and re-covered by Jean Salis Sr, the Musée de l'Air example is now on display at Le Bourget. Here it is at the Chalais Meudon store. (R Moulton)
▼

▲ **Capt Eddie Rickenbacker poses with his famous SPAD XIII from the 94th Pursuit Sqn. He scored 25 victories. (Air Photos 138)**

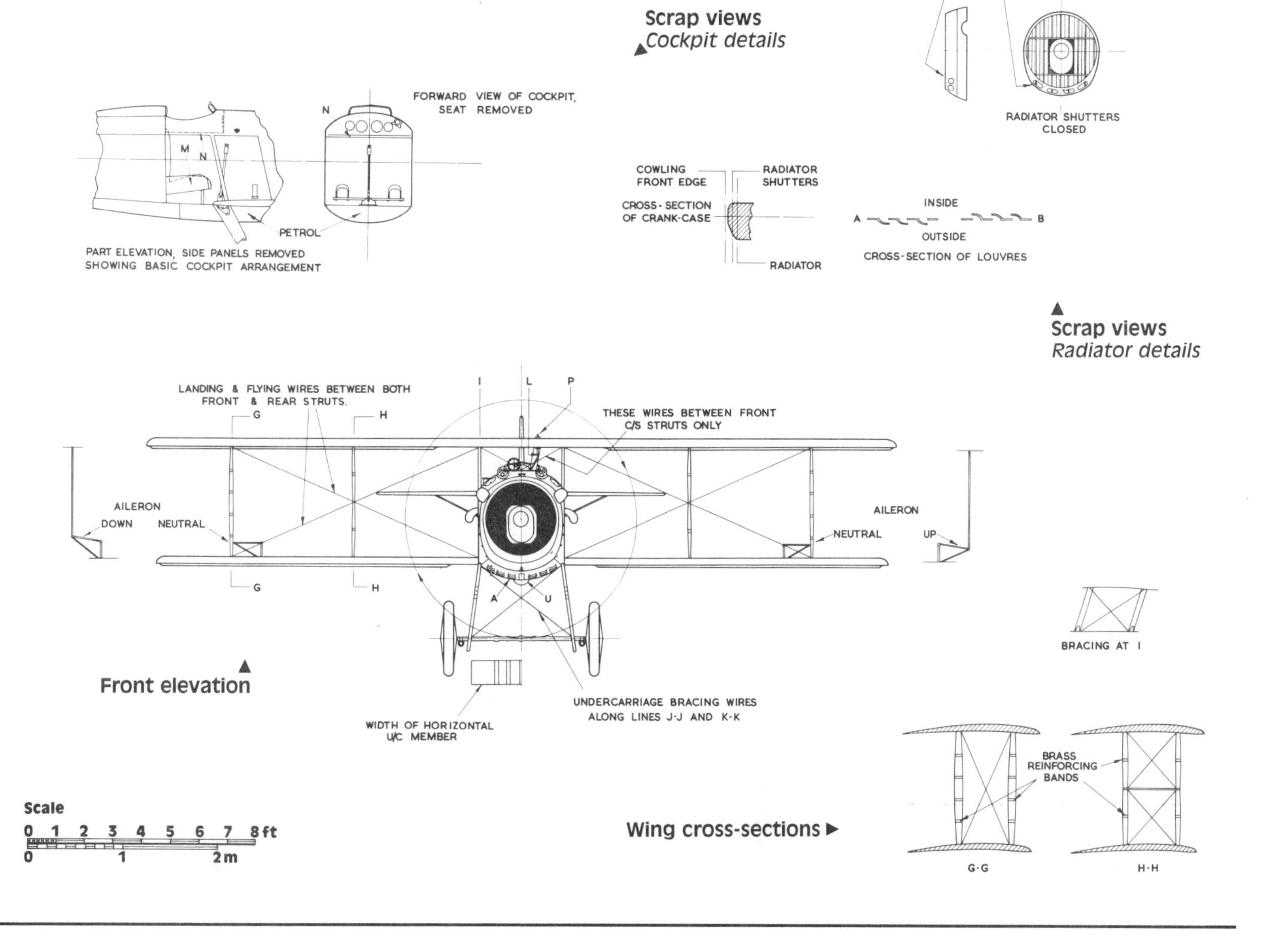

Scrap views
▲*Cockpit details*

▲
Scrap views
Radiator details

▲
Front elevation

Wing cross-sections ▶

Thomas-Morse S-4C

Country of origin: USA.
Type: Single-seat trainer.
Powerplant: One Le Rhône rotary engine rated at 80hp or Gnome rated at 100hp.
Dimensions: Wing span 26ft 6in *8.08m*; length 19ft 10in *6.05m*; height 8ft 1in *2.46m*; wing area 145 sq ft *13.5m²*.
Weights: Gross 1330lb *603kg*.
Performance: Maximum speed 97mph *156kph*; time to 7500ft *2285m*, 10min.
Armament: One fixed Marlin 0.3in machine gun.
Service: First flight (prototype) June 1917.

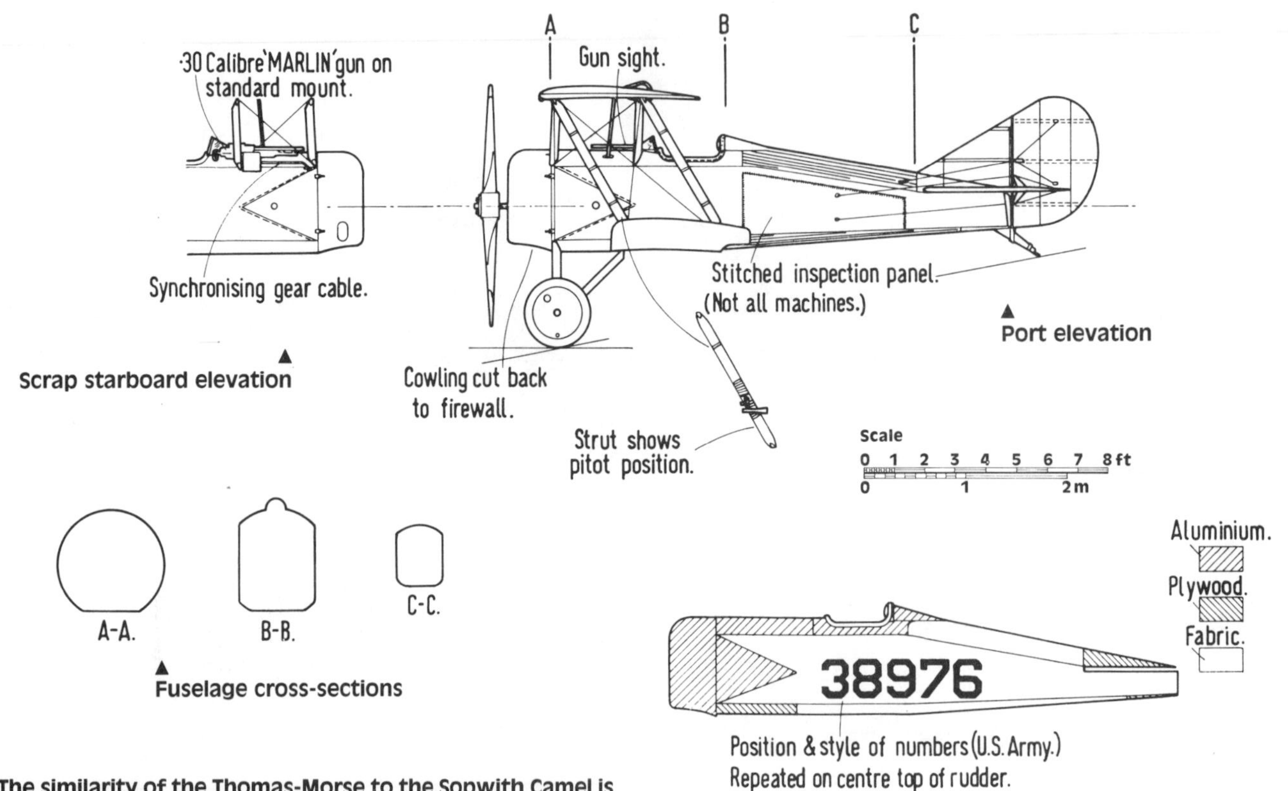

The similarity of the Thomas-Morse to the Sopwith Camel is no coincidence as both had a common ancestor in the Sopwith Tabloid. B Douglas Thomas, the designer, had worked for Vickers and Sopwith before his emigration to the USA in 1915. (USAF)

▲
Fuselages await their 80hp Le Rhône engines at the Thomas Morse plant. Stringered upper coamings and cockpit fairings are another separate assembly.

First flown in June 1917, the Thomas-Morse S-4 was improved through the S-4B to the eventual S-4C. 'Tommies' were produced at Ithica, New York State, to orders totalling 400 from the US War Department; postwar, many were operated privately, and for films such as *Dawn Patrol.*
▼

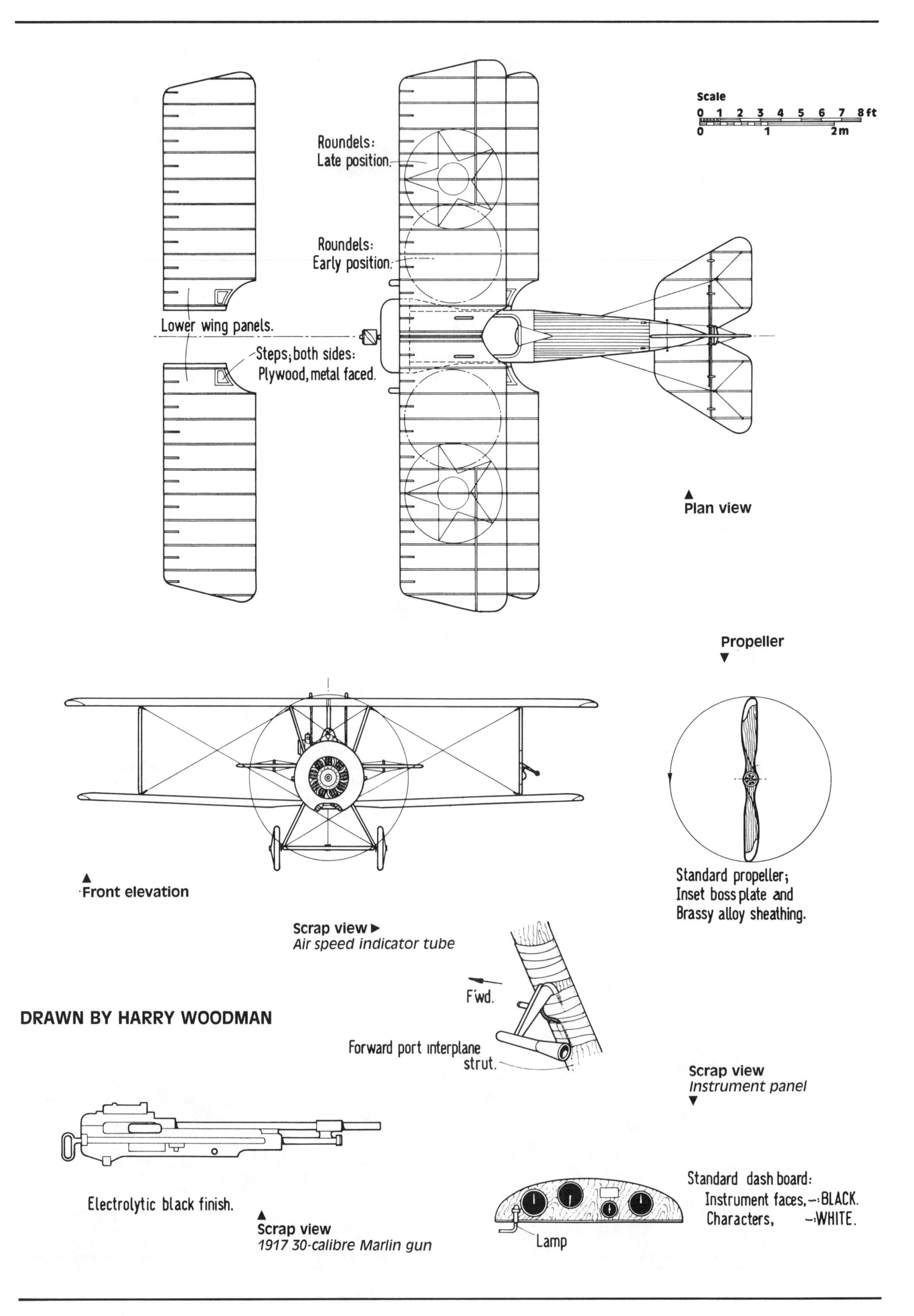

DRAWN BY HARRY WOODMAN

Albatros C III

Country of origin: Germany.
Type: Two-seat general-purpose aircraft.
Powerplant: One Benz Bz III engine rated at 150hp or Mercedes D III rated at 160hp.
Dimensions: Wing span 38ft 3in *11.69m*; length 26ft 3in *8.00m*; height 10ft 0¾in *3.07m*, (D III) 10ft 2in *3.10m*; wing area 397 sq ft *36.91m²*.
Weights: Empty 1876lb *851kg*; loaded 2983lb *1353kg*.
Performance: Maximum speed 87mph *140kph*; time to 3280ft *1000m*, 9min; service ceiling 11,000ft *3350m*; endurance 4hr.
Armament: One flexibly mounted Parabellum machine gun and (later) one fixed machine gun, plus 200lb *90kg* of bombs.
Service: Service entry 1916.

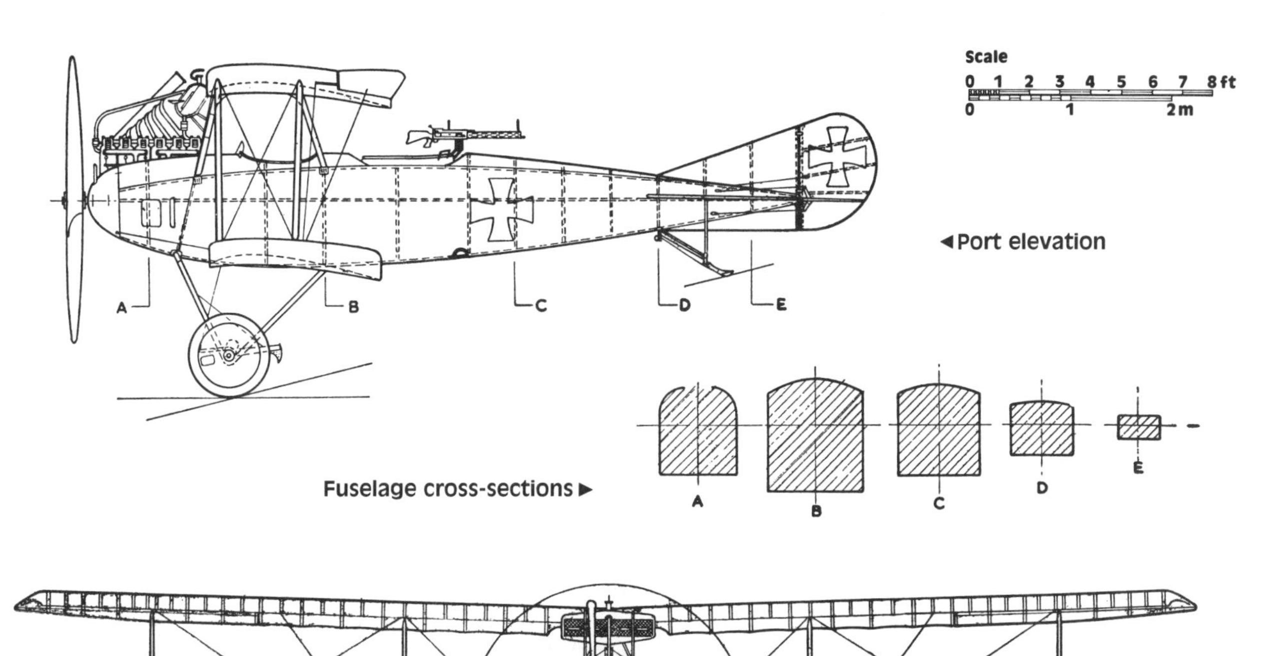

◄Port elevation

Fuselage cross-sections ►

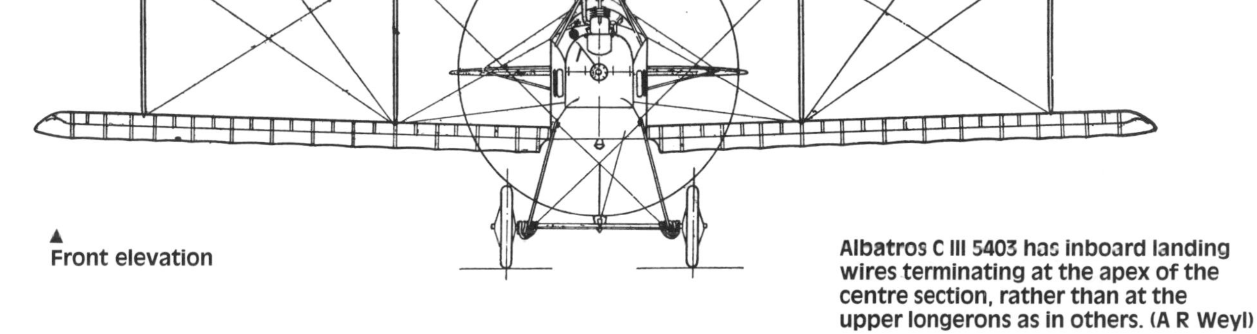

▲ Front elevation

Albatros C III 5403 has inboard landing wires terminating at the apex of the centre section, rather than at the upper longerons as in others. (A R Weyl) ▼

Colour notes

Early C IIIs were not camouflaged, being covered with varnished linen fabric. Patee crosses were painted above the top and under the lower wings, on both sides of the fuselage and rudder, sometimes against a white square. With the introduction of camouflage, the upper and side surfaces were painted mainly in a 'shadow-shaded' form in tones of dark olive green and dark (though not dull) purple. Undersurfaces were either plain varnished linen fabric or overpainted in a pale sky blue colour. Serial numbers were black, usually along the centreline of the rear fuselage in either a plain block style or a slightly ornate Roman style.

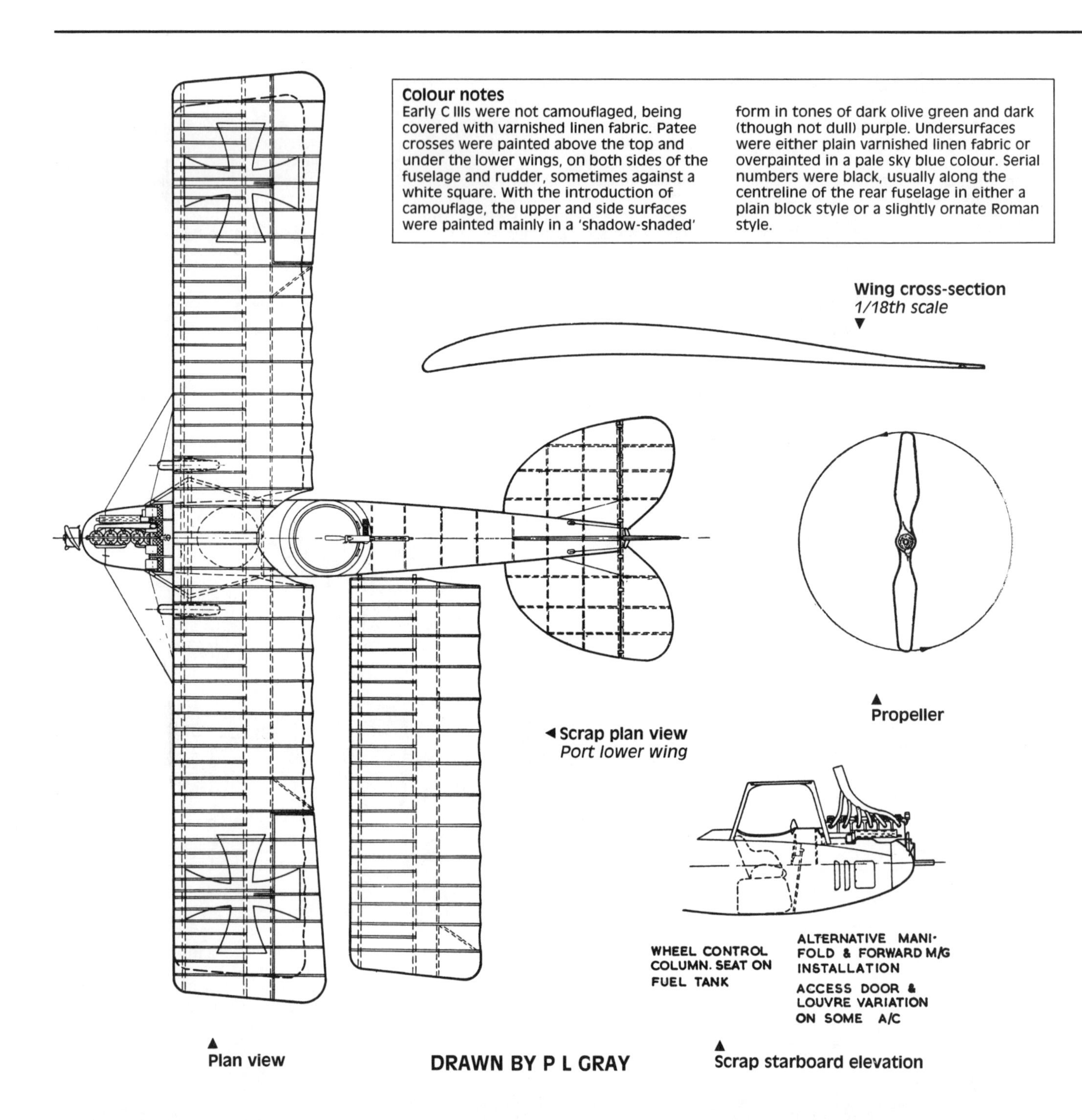

Plan view

DRAWN BY P L GRAY

Scrap starboard elevation

Exhaust manifold on a Hofmann-built C III. Unpainted finish and absence of forward gun suggest that this aircraft was non-operational. (A R Weyl)

A Hansa-built C III with the radiator shutters removed. This useful port-side view shows small detail from cowling retainers to carburettor manifold. (A R Weyl)

▲▲
Gun and interrupter gear are revealed in this close-up of a C III captured in Macedonia. (G Haddow)

▲
Cockpit of late Albatros C X differed from the C III in having a four-spoked control wheel. Oxygen equipment is in left foreground. Rudder bar is short! (A R Weyl)

▲
Radio equipment in a C X is comprehensive and the rear cockpit better fitted than the C III. Engine controls appear to be duplicated. (A R Weyl)

An Albatros C III (Li) with its wheel discs and clam brake removed. The low light angle, directly head-on, emphasises the undercambered aerofoil of both wings. (A R Weyl)
▼

Albatros D II

Country of origin: Germany.
Type: Single-seat scout.
Powerplant: One Mercedes D IIIa six-cylinder, liquid-cooled engine rated at 160hp.
Dimensions: Wing span 27ft 10½in *8.50m*; length 24ft 3in *7.40m*; height 8ft 8in *2.64m*.
Weights: Empty 1405lb *637kg*; loaded 1958lb *888kg*.
Performance: Maximum speed 109mph *175kph*; time to 3280ft *1000m*, 5min; endurance 1.5hr.
Armament: Two fixed Spandau machine guns.
Service: Service entry autumn 1916.

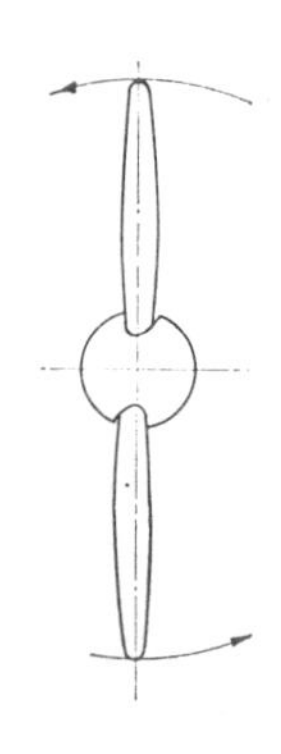

▲ Propeller

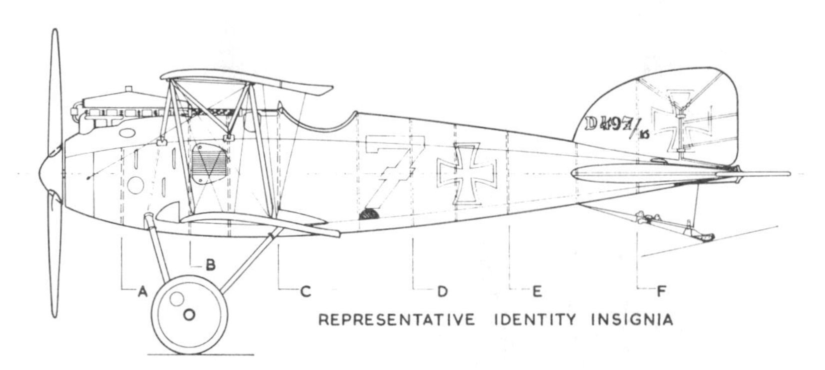

▲ Port elevation, D II

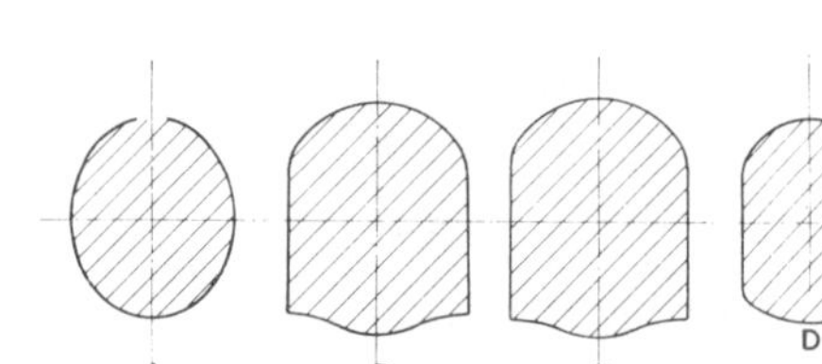
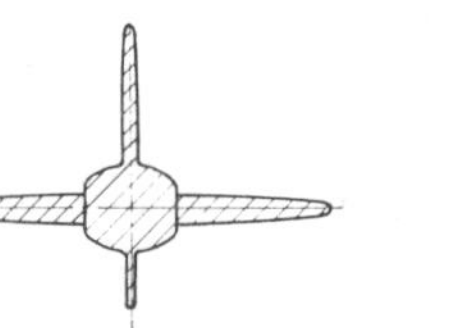

▲ Fuselage cross-sections

DRAWN BY P L GRAY

A 180hp Austro-Daimler powered Albatros D II. It was in a 160hp Mercedes D II that Manfred von Richthofen scored his first victory, his eleventh being the DH2 of Major Lanoe Hawker. Oswald Boelcke achieved 40 victories, many in the D II while commanding *Jasta 2.* (H J Nowarra) ▼

Colour notes

Fabric surfaces (wings, tailplane, control surfaces) were camouflaged in large, irregular patches of dark olive green and deep mauve on the upper surfaces, the wing surface being divided into three or four patches. Undersurfaces were usually doped with clear sky blue. Ex-works, the fuselage was generally left in natural finish, with several coats of protective varnish, which imparted a yellowish (straw) shade, though some units painted the fuselage dark olive. Steel struts and metal nose panels were painted for protection either grey or dark green. Patee crosses were at first painted on square white backgrounds but later with only a white outline. At the period the D I and D II were operational (circa autumn 1916 to spring 1919) the grey colour schemes of the *Jagdstaffeln* had not yet come into usage. It was usually found that a large numeral or personal device such as a monogram, swastika etc painted on the fuselage sides in a contrasting colour was sufficient to identify a comrade in the air. Later some *Jastas* also adopted coloured tails as a unit identity: Boelcke's *Jasta 2* had black and white tailplanes, starboard half black and port half white, on its machines.

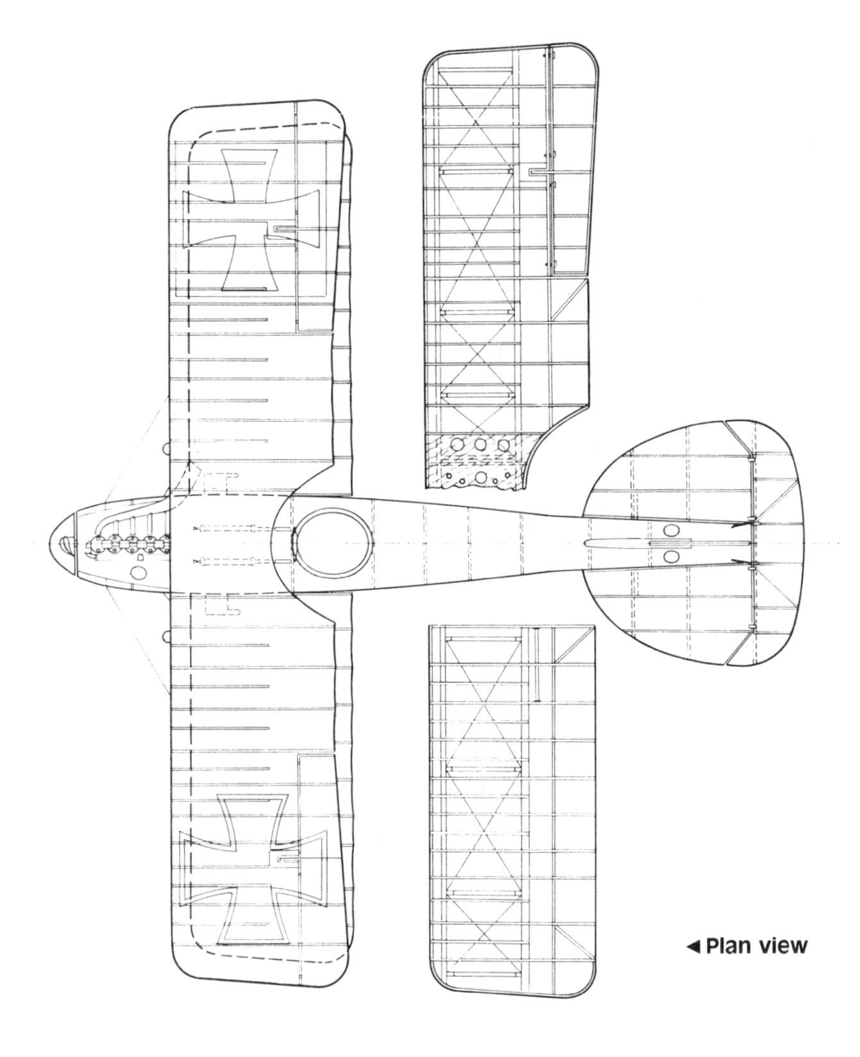

◄ Plan view

Scale

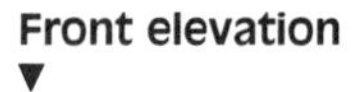

Front elevation ▼

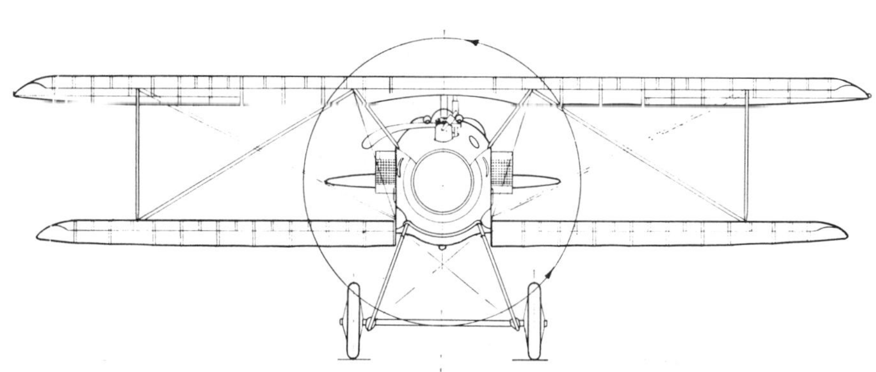

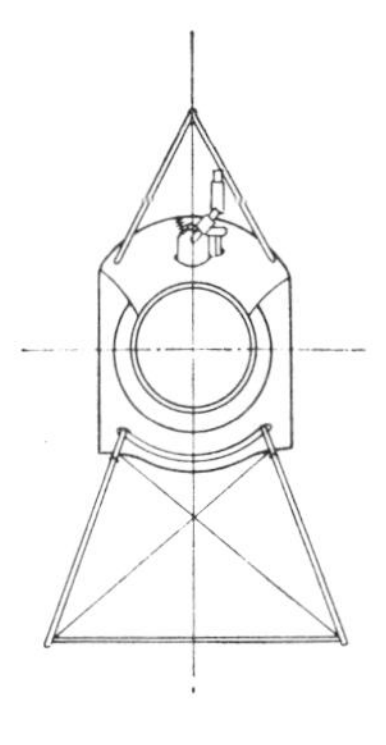

▲ Scrap front elevation, D I

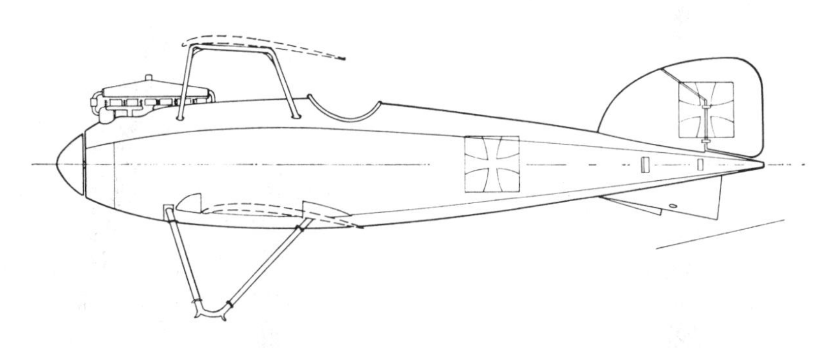

Port elevation, D I ►
Showing trestle-type centre-section cabane and increased gap

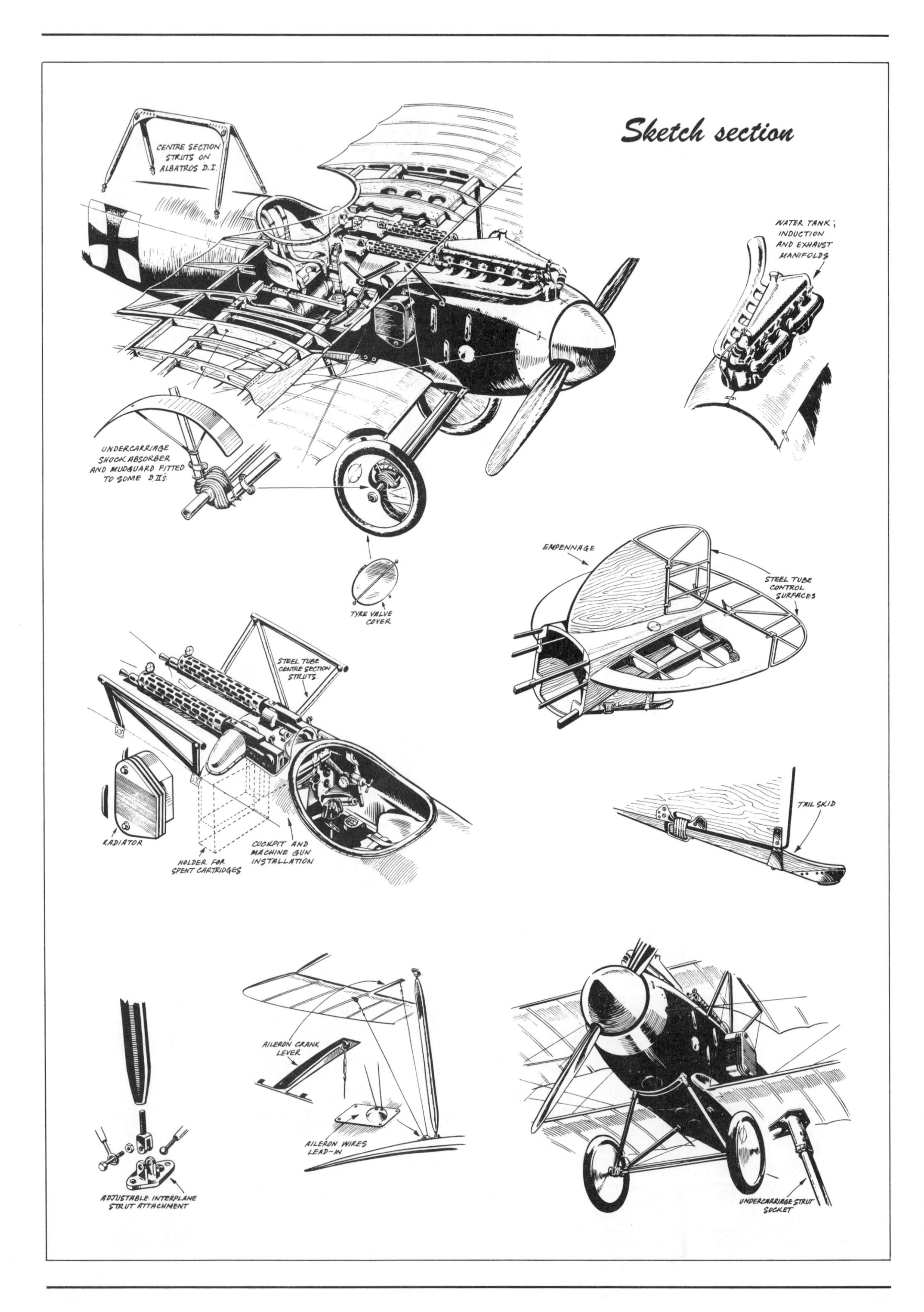

Sketch section
CENTRE SECTION STRUTS ON ALBATROS D.I.
WATER TANK; INDUCTION AND EXHAUST MANIFOLDS
UNDERCARRIAGE SHOCK ABSORBER AND MUDGUARD FITTED TO SOME D.II's
TYRE VALVE COVER
EMPENNAGE
STEEL TUBE CONTROL SURFACES
STEEL TUBE CENTRE SECTION STRUTS
RADIATOR
HOLDER FOR SPENT CARTRIDGES
COCKPIT AND MACHINE GUN INSTALLATION
TAIL SKID
AILERON CRANK LEVER
AILERON WIRES LEAD-IN
ADJUSTABLE INTERPLANE STRUT ATTACHMENT
UNDERCARRIAGE STRUT SOCKET

Albatros D V and Va

Country of origin: Germany.
Type: Single-seat scout.
Powerplant: One Mercedes D IIIa six-cylinder, liquid-cooled engine rated at 180hp or 200hp.
Dimensions: Wing span 29ft 8¼in *9.05m*; length 24ft 0½in *7.33m*; height 8ft 10¼in *2.70m*; wing area 229 sq ft *21.3m²*.
Weights: Empty 1515lb *687kg*; loaded 2066lb *937kg*.
Performance: Maximum speed 115.5mph *186kph*; time to 3280ft *1000m*, 4min; service ceiling 20,000ft *6100m*; endurance 2hr.
Armament: Two fixed Spandau machine guns.
Service: Service entry summer 1917.

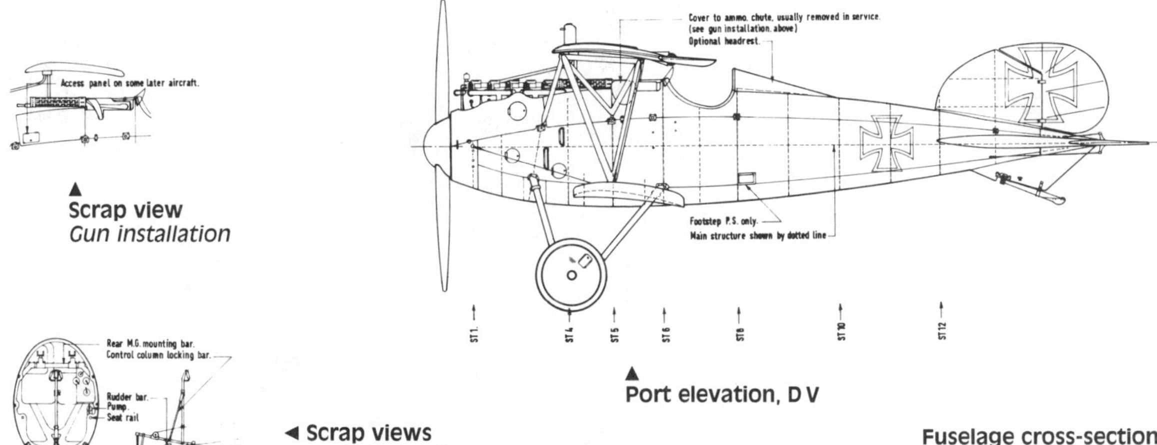

▲ Scrap view
Gun installation

▲ Port elevation, D V

Rear M.G. mounting bar.
Control column locking bar.
Rudder bar.
Pump.
Seat rail
SECTION ST6.
CONTROL DETAIL

◄ Scrap views
Cockpit details

Fuselage cross-sections
▼

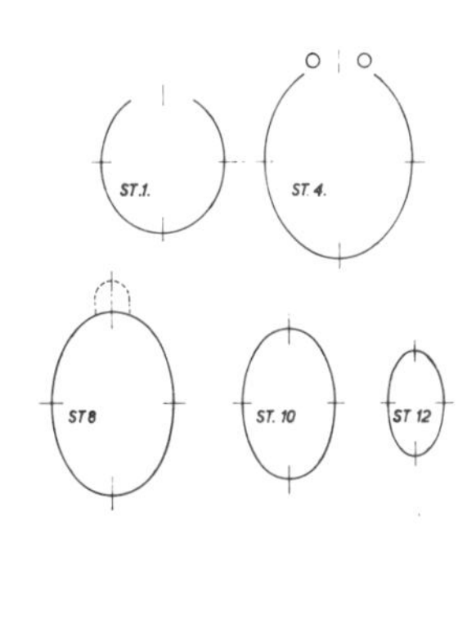

Rudder fabric covered
Fin plywood covered, part of fuselage.
Wire strap to hold down cowling – optional.
Control cables.
Extra strut on late model D.Va's.

▲ Starboard elevation, D Va

Prototype D V, its plywood fuselage hand-painted with the lozenge pattern camouflage to match the wing fabric. (P L Gray)
▼

▲
Force-landed in Switzerland, this D Va is on show under internment. (E van Gorder)

DRAWN BY IAN R STAIR

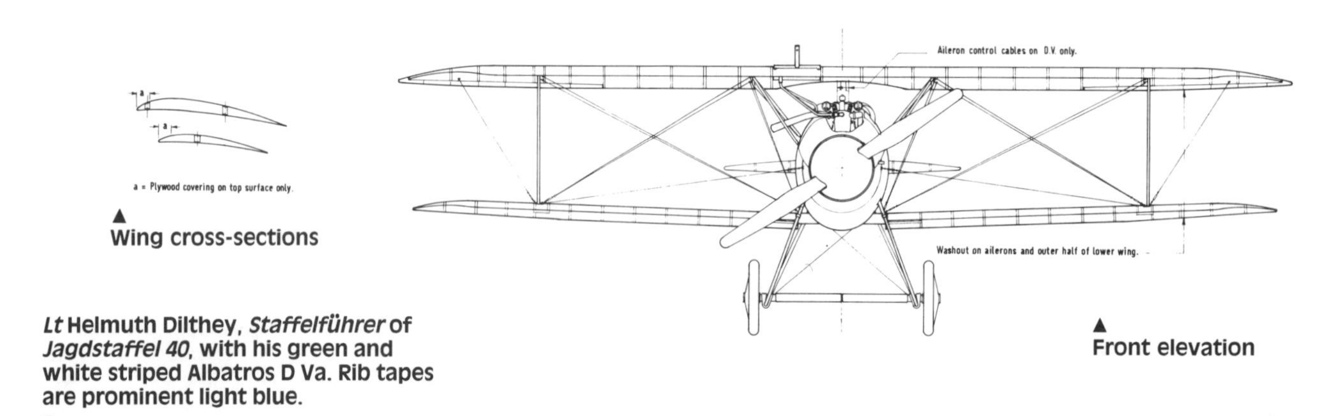

▲
Wing cross-sections

▲
Front elevation

Lt Helmuth Dilthey, *Staffelführer* of *Jagdstaffel 40*, with his green and white striped Albatros D Va. Rib tapes are prominent light blue.
▼

▲ Albatros D Va D5390/17, captured on 7 December 1917 by Australians, is now displayed, after restoration, in the National War Memorial, Canberra. (IWM E1685)

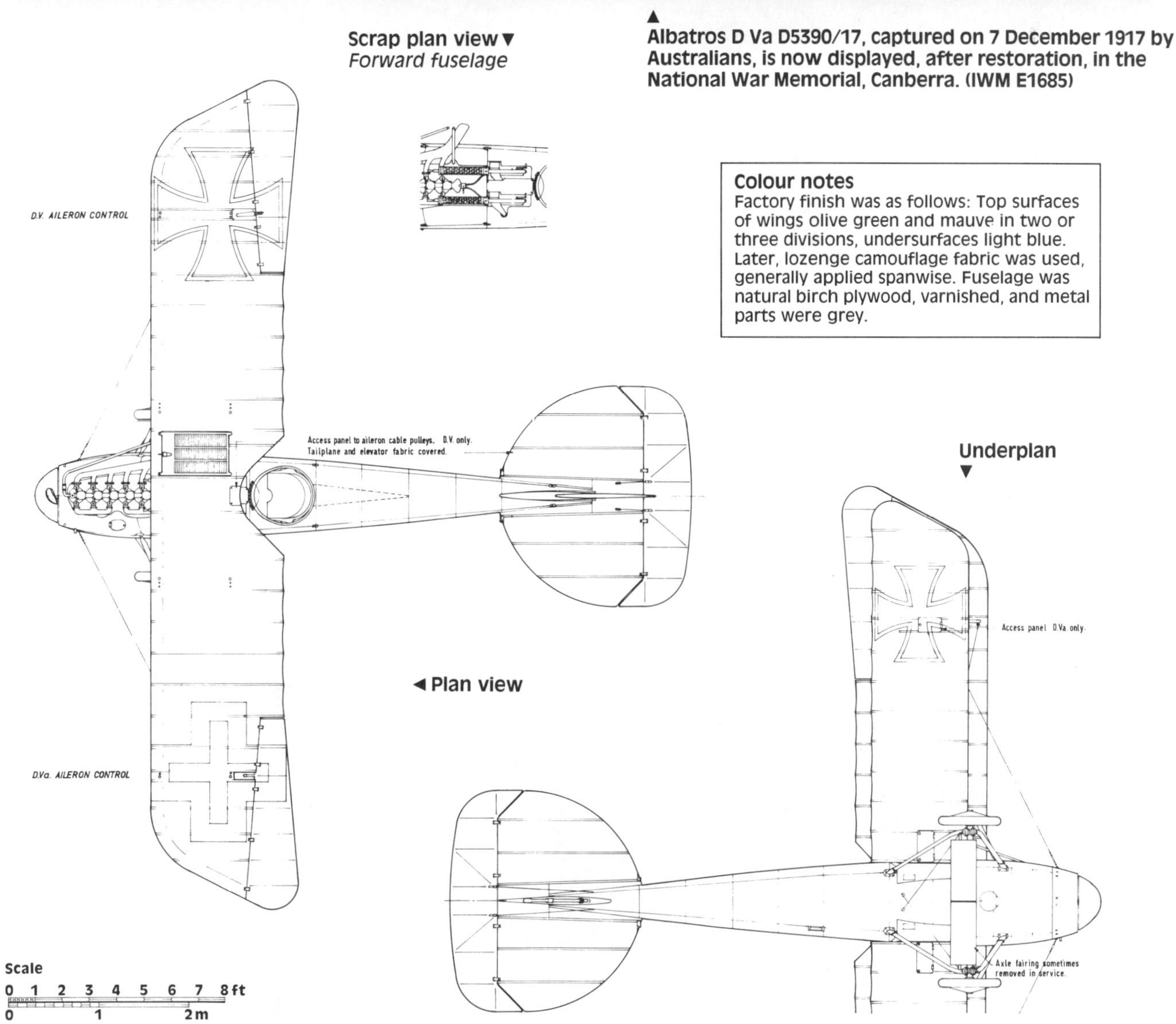

Colour notes

Factory finish was as follows: Top surfaces of wings olive green and mauve in two or three divisions, undersurfaces light blue. Later, lozenge camouflage fabric was used, generally applied spanwise. Fuselage was natural birch plywood, varnished, and metal parts were grey.

Fokker D VIII

Country of origin: Germany.
Type: Single-seat scout.
Powerplant: One Oberursel U II nine-cylinder rotary engine rated at 110hp.
Dimensions: Wing span 27ft 4¼in *8.34m*; length 19ft 2¾in *5.86m*; height 8ft 6½in *2.60m*; wing area 115.5 sq ft *10.7m²*.
Weights: Empty 893lb *405kg*; loaded 1334lb *605kg*.
Performance: Maximum speed 127mph *204kph* at ground level; time to 3280ft *1000m*, 2min; service ceiling 19,700ft *6000m*; endurance 1.5hr.
Armament: Two fixed Spandau machine guns.
Service: Service entry (E Ve) July 1918, (D VIII) September 1918.

DRAWN BY IAN R STAIR

Port elevation ►

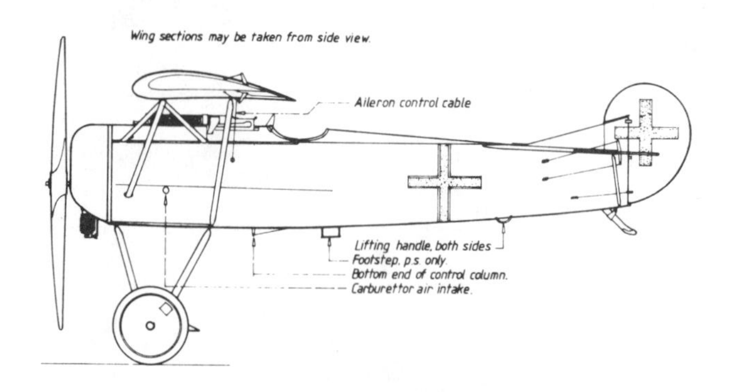

Scale
0 1 2 3 4 5 6 7 8 ft
0 1 2 m

Starboard inboard profile
▼

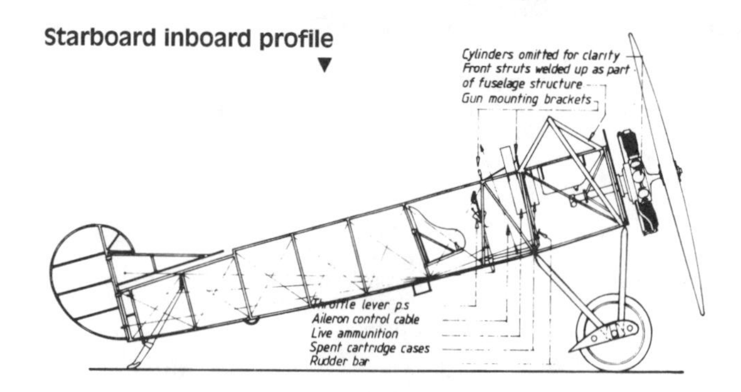

Colour notes
Wing, axle wing, cowling and struts – dark green; fabric covering – lozenge camouflage. Rudder and aft part of fin – white. Crosses – black with white edges on wing and fuselage.

Fokker V28, an unarmed prototype of the D VIII fitted with 200hp Goebel engine and devoid of miltary markings. Although obscure, this perfect side elevation is a useful reference for modellers. (H J Nowarra)
▼

Ernst Udet in his Fokker D VIII 238, with lozenge-pattern fabric on the fuselage. (E van Gorder) ►

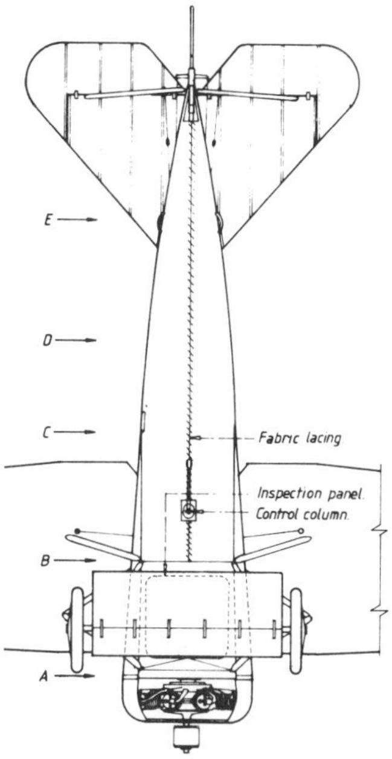

Fabric lacing
Inspection panel
Control column

▲
Underplan

Scrap plan view
Detail beneath top wing
▼

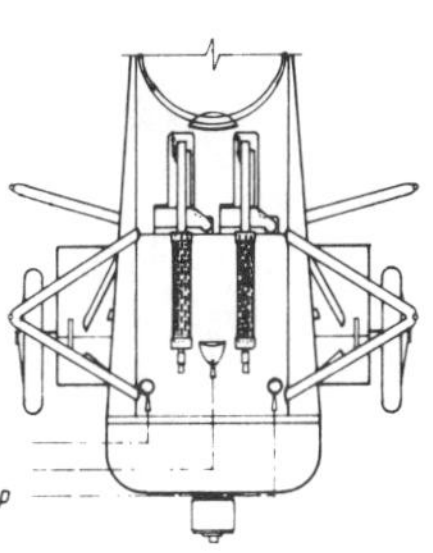

Oil tank filler cap
Fuel gauge cover
Fuel tank filler cap

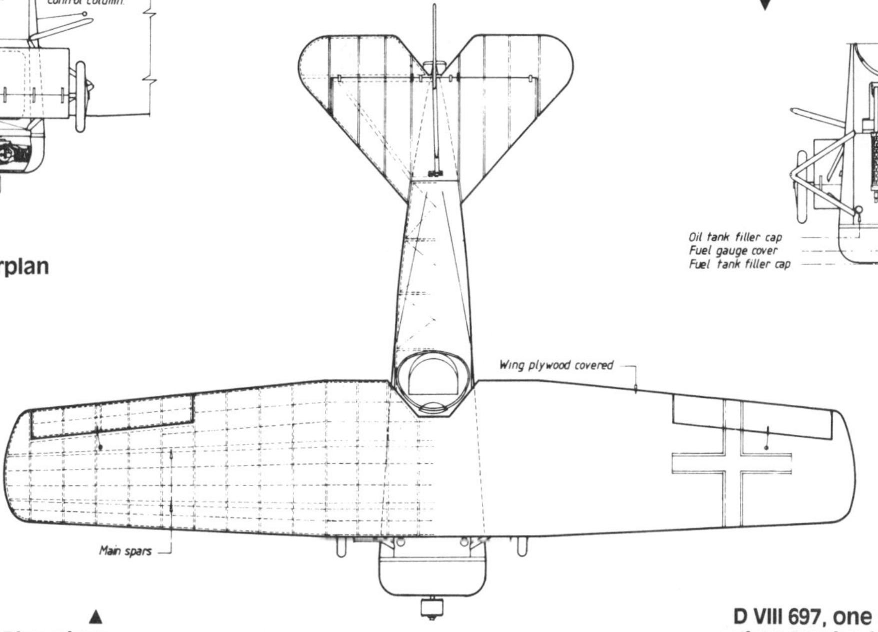

Wing plywood covered
Main spars

▲
Plan view

D VIII 697, one of the later series. This view emphasises the centre-section cutaway. (H J Nowarra)
▼

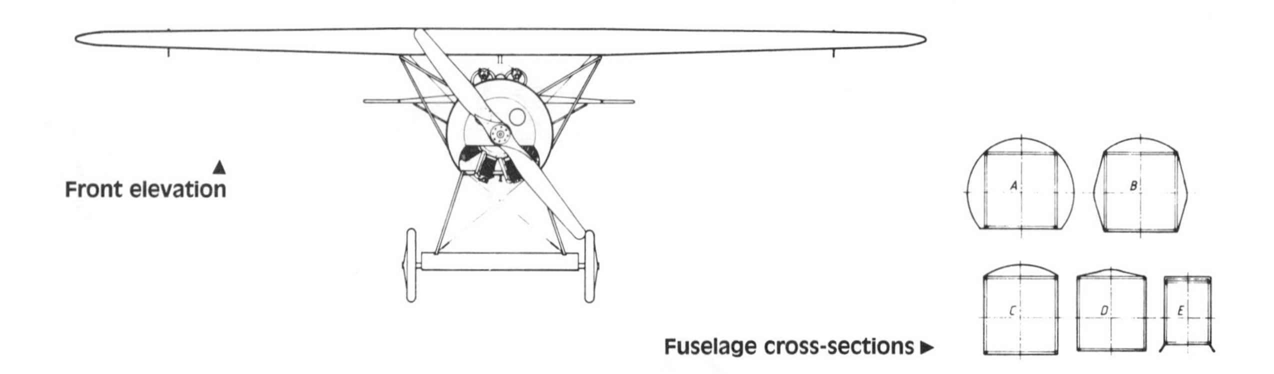

▲ Front elevation

Fuselage cross-sections ▶

◀ One of the first Fokker E Vs, 138/18, which served with the German naval fighter units in Flanders during August 1918. (H J Nowarra)

Dark but useful reference view of late D VIII 697 (see also rear three-quarter view), with the Spandau guns prominent under the centre section. (H J Nowarra)
▼

Fokker E III and IV

Country of origin: Germany.
Type: Single-seat scout.
Powerplant: One Oberursel Ur I nine-cylinder rotary engine rated at 100hp, (E IV) Oberursel Ur III rated at 160hp.
Dimensions: Wing span 31ft 2¾in *9.52m*; length 23ft 7½in *7.20m*; height 7ft 10½in *2.40m*; wing area 172.8 sq ft *16.0m²*.
Weights: Empty 880lb *399kg*; loaded (E III) 1345lb *610kg*.
Performance: Maximum speed (E III) 87mph *140kph*; time to 3280ft *1000m*, 5min; service ceiling 11,500ft *3500m*; endurance 1.5hr.
Armament: One fixed Spandau machine gun.
Service: First flight (E III) late 1915, (E IV) spring 1916.

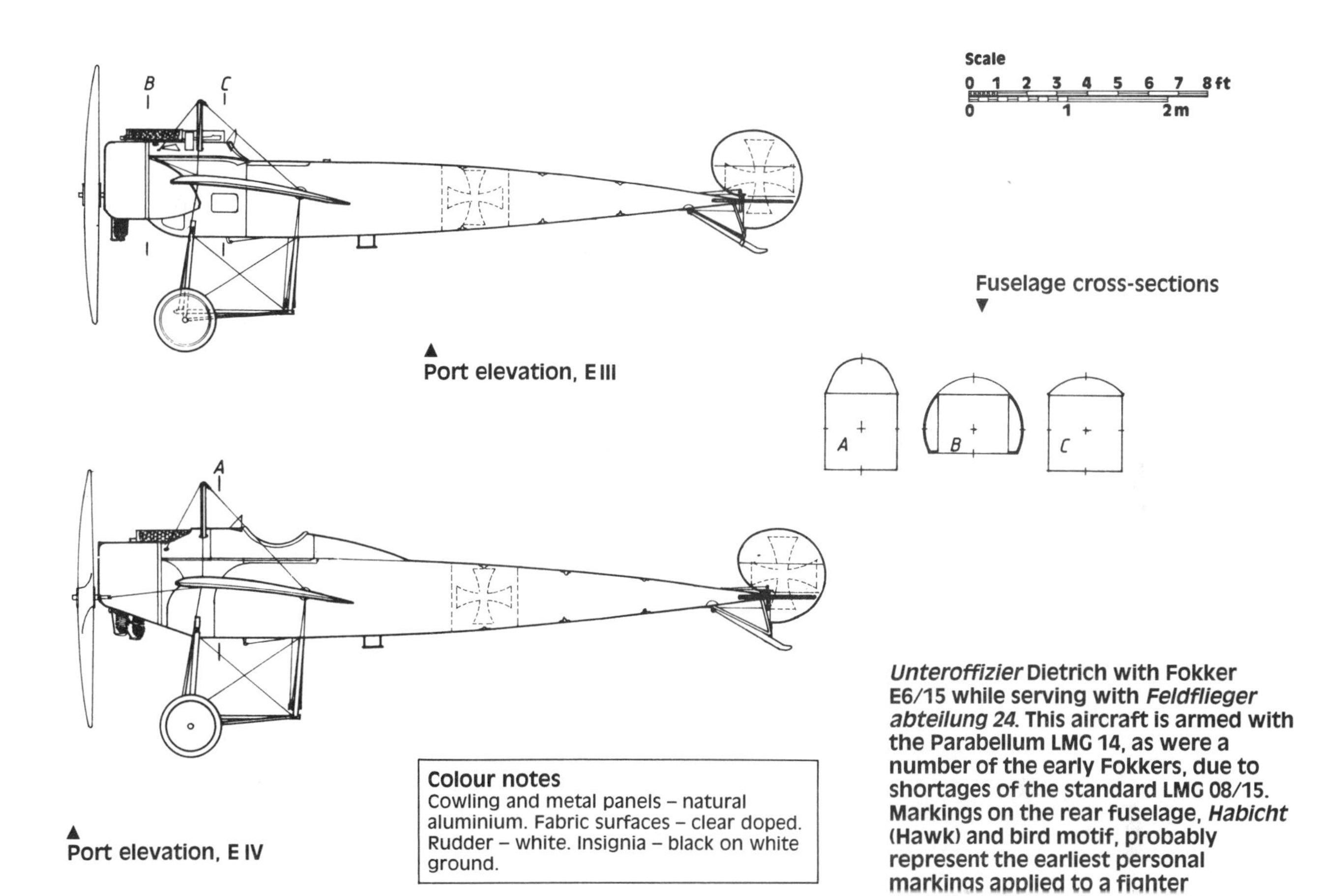

Port elevation, E III

Port elevation, E IV

Colour notes
Cowling and metal panels – natural aluminium. Fabric surfaces – clear doped. Rudder – white. Insignia – black on white ground.

***Unteroffizier* Dietrich with Fokker E6/15 while serving with *Feldflieger abteilung 24*. This aircraft is armed with the Parabellum LMG 14, as were a number of the early Fokkers, due to shortages of the standard LMG 08/15. Markings on the rear fuselage, *Habicht* (Hawk) and bird motif, probably represent the earliest personal markings applied to a fighter aeroplane. (A Imrie)**

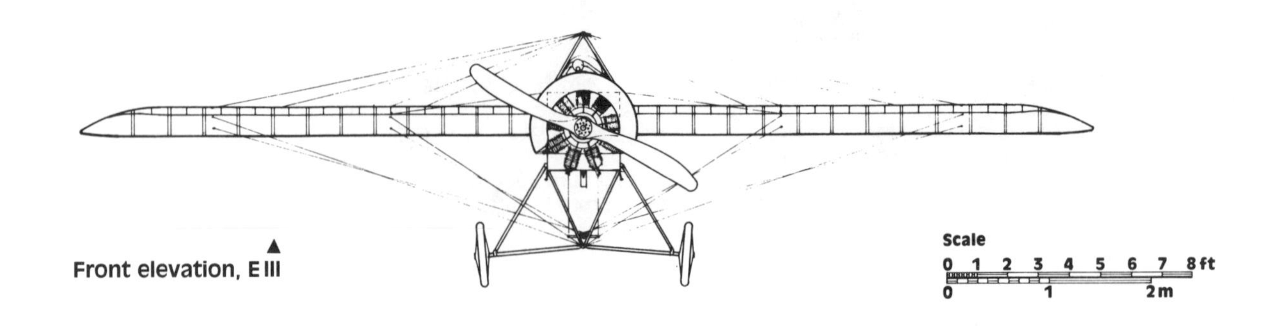

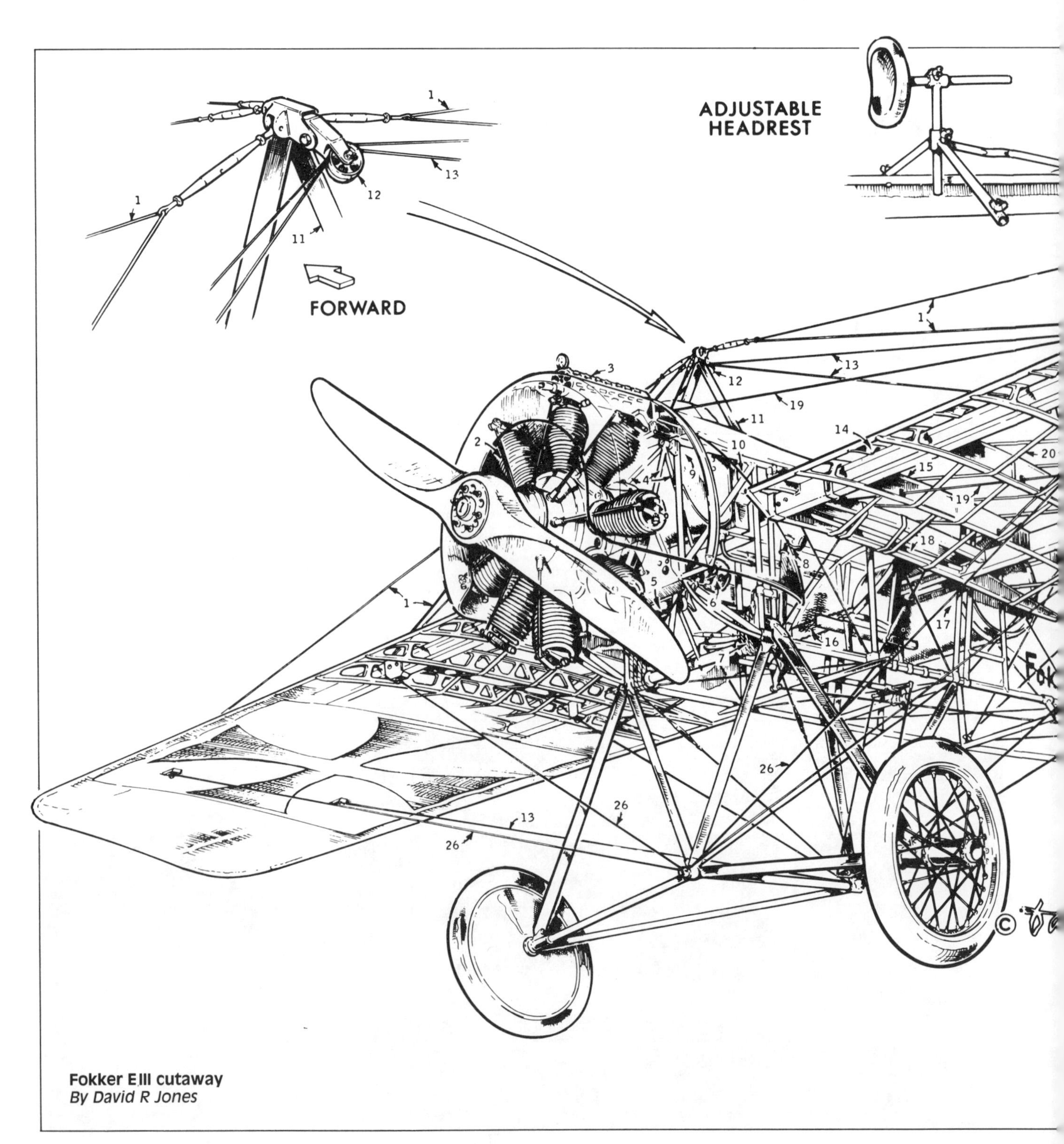

Fokker E III cutaway
By David R Jones

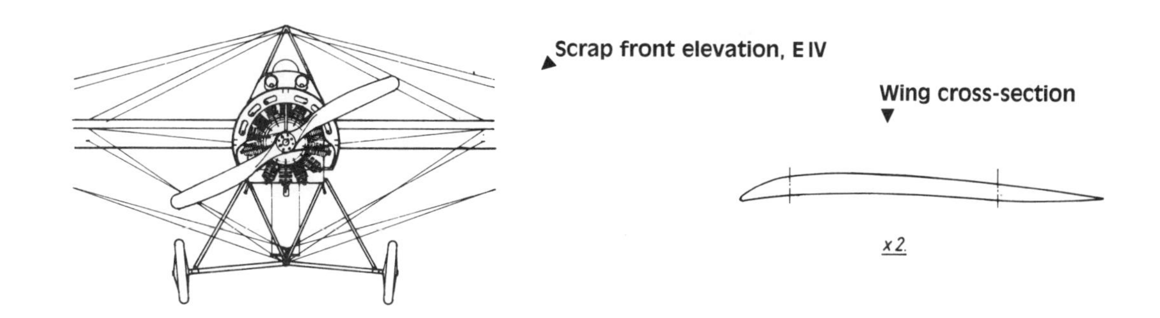

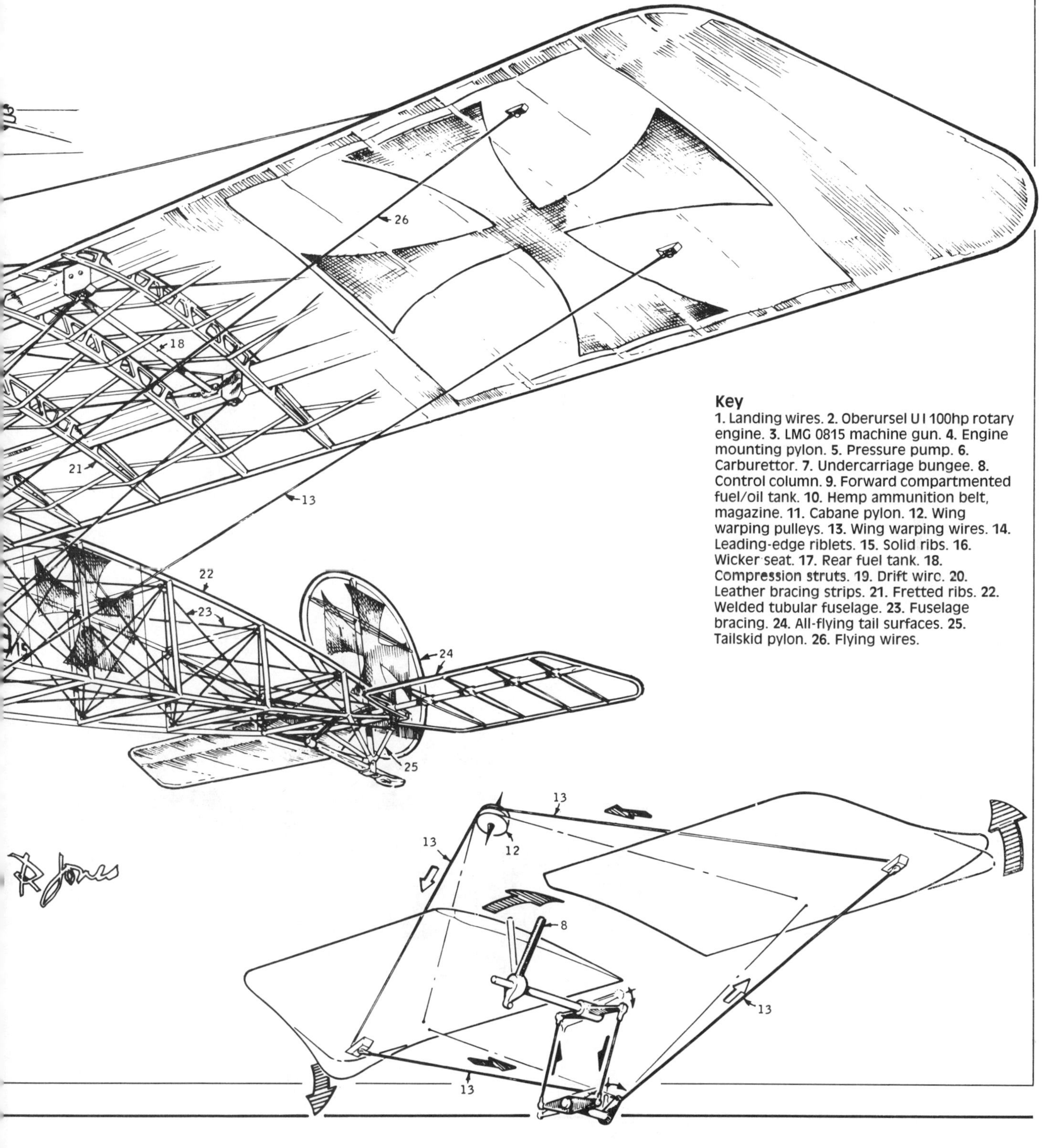

Key
1. Landing wires. **2.** Oberursel U I 100hp rotary engine. **3.** LMG 0815 machine gun. **4.** Engine mounting pylon. **5.** Pressure pump. **6.** Carburettor. **7.** Undercarriage bungee. **8.** Control column. **9.** Forward compartmented fuel/oil tank. **10.** Hemp ammunition belt, magazine. **11.** Cabane pylon. **12.** Wing warping pulleys. **13.** Wing warping wires. **14.** Leading-edge riblets. **15.** Solid ribs. **16.** Wicker seat. **17.** Rear fuel tank. **18.** Compression struts. **19.** Drift wirc. **20.** Leather bracing strips. **21.** Fretted ribs. **22.** Welded tubular fuselage. **23.** Fuselage bracing. **24.** All-flying tail surfaces. **25.** Tailskid pylon. **26.** Flying wires.

DRAWN BY IAN R STAIR

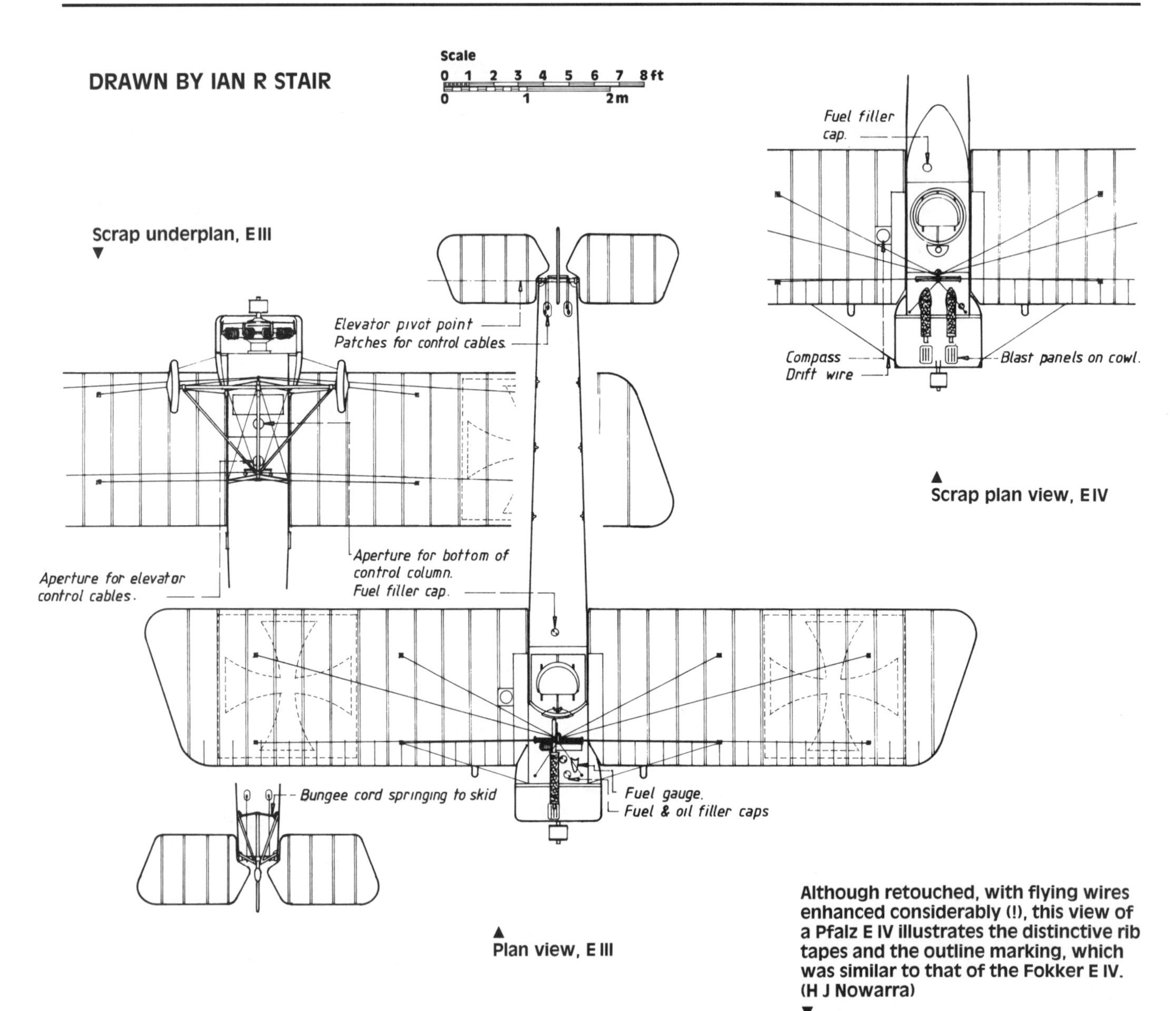

▲ Scrap plan view, E IV

▲ Plan view, E III

Although retouched, with flying wires enhanced considerably (!), this view of a Pfalz E IV illustrates the distinctive rib tapes and the outline marking, which was similar to that of the Fokker E IV. (H J Nowarra)
▼

Halberstadt CL II

Country of origin: Germany.
Type: Two-seat fighter and close-support aircraft.
Powerplant: One Mercedes D III six-cylinder, liquid-cooled engine rated at 160hp, (later aircraft) Mercedes D IIIa rated at 180hp.
Dimensions: Wing span 35ft 4in *10.77m*; length 23ft 11½in *7.30m*; height 9ft 0½in *2.75m*.
Weights: Empty 1704lb *773kg*; loaded 2498lb *1133kg*.
Performance: Maximum speed 102.5mph *165kph* at 16,400ft *5000m*; time to 3280ft *1000m*, 5min; service ceiling 16,700ft *5100m*; endurance about 3hr.
Armament: One or two fixed Spandau machine guns and one flexibly mounted Parabellum machine gun.
Service: Service entry late summer 1917.

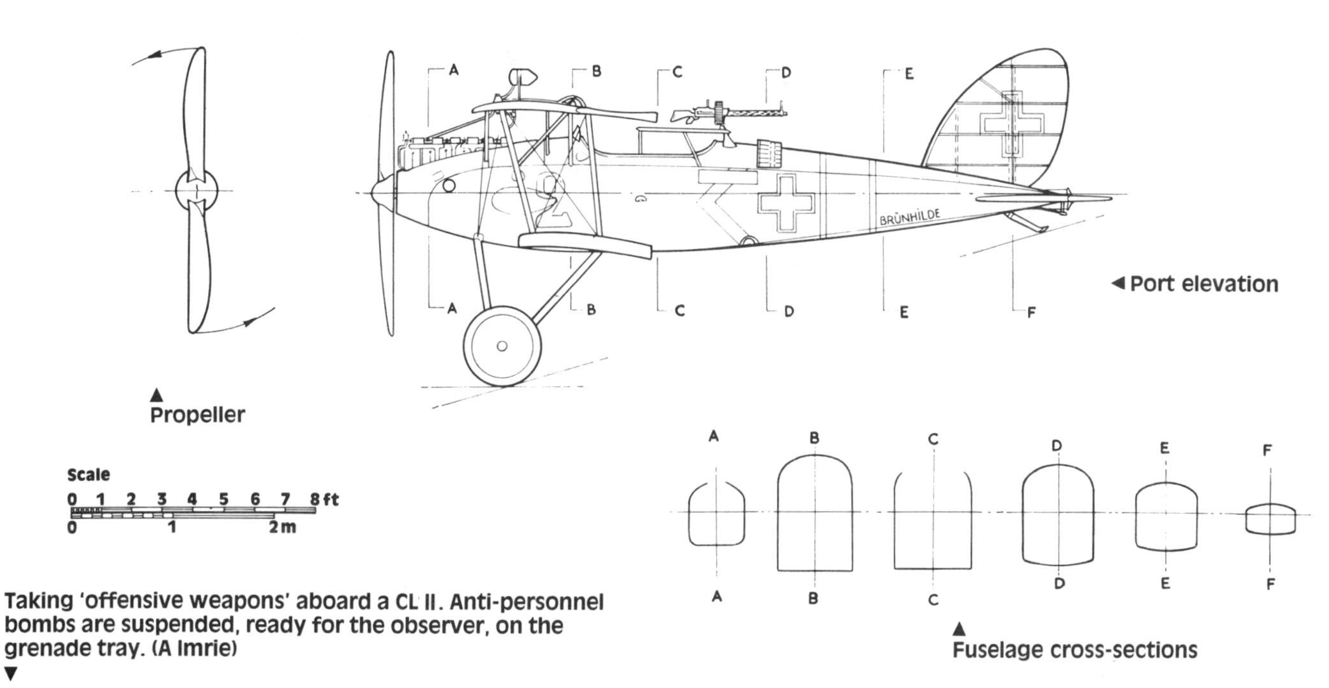

▲ Propeller

◄ Port elevation

▲ Fuselage cross-sections

Taking 'offensive weapons' aboard a CL II. Anti-personnel bombs are suspended, ready for the observer, on the grenade tray. (A Imrie)
▼

▲
On type-test at Adlershof in May 1917, this Halberstadt CL II has the contrasting laminated timber airscrew on its Mercedes engine and shows its horn-balanced ailerons.

DRAWN BY P L GRAY

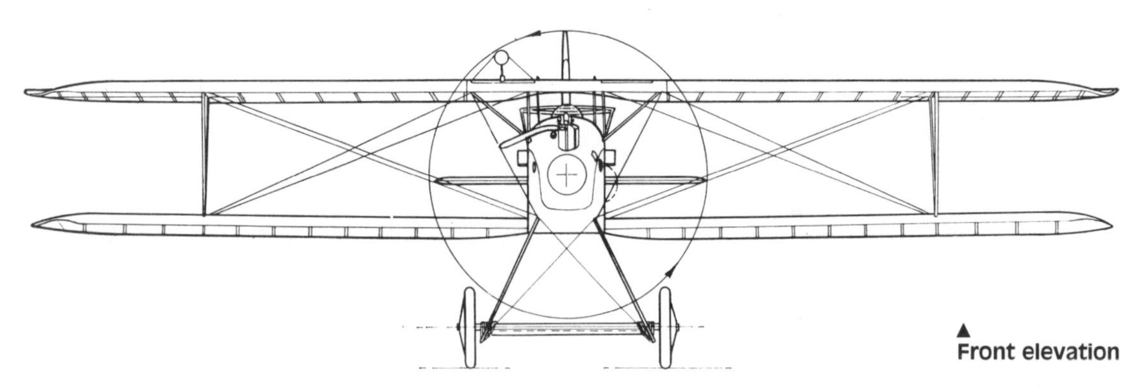

▲
Front elevation

The communal cockpit for pilot and observer makes a group pose easy for this post-Armistice photograph.
(E Kreuger)
▼

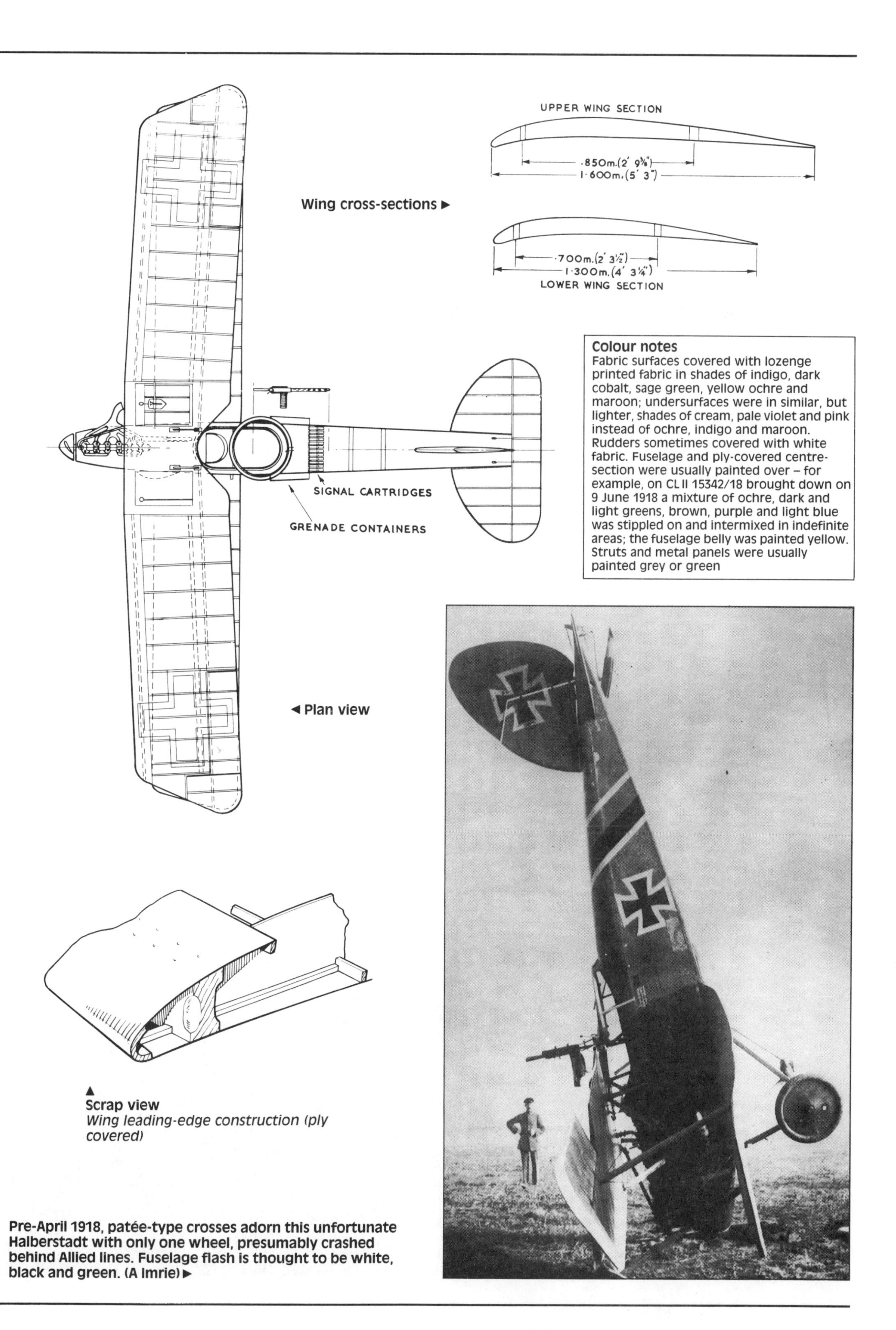

Wing cross-sections ▶

◀ Plan view

Colour notes
Fabric surfaces covered with lozenge printed fabric in shades of indigo, dark cobalt, sage green, yellow ochre and maroon; undersurfaces were in similar, but lighter, shades of cream, pale violet and pink instead of ochre, indigo and maroon. Rudders sometimes covered with white fabric. Fuselage and ply-covered centre-section were usually painted over – for example, on CL II 15342/18 brought down on 9 June 1918 a mixture of ochre, dark and light greens, brown, purple and light blue was stippled on and intermixed in indefinite areas; the fuselage belly was painted yellow. Struts and metal panels were usually painted grey or green

▲
Scrap view
Wing leading-edge construction (ply covered)

Pre-April 1918, patée-type crosses adorn this unfortunate Halberstadt with only one wheel, presumably crashed behind Allied lines. Fuselage flash is thought to be white, black and green. (A Imrie) ▶

Hannover CL IIIa

Country of origin: Germany.
Type: Two-seat light escort and close-support aircraft.
Powerplant: One Argus As III six-cylinder, liquid-cooled engine rated at 180hp.
Dimensions: Wing span 38ft 4¾in *11.70m*; length 24ft 10½in *7.58m*; height 9ft 2¼in *2.80m*; wing area 353 sq ft *32.7m²*.
Weights: Empty 1581lb *717kg*; loaded 2383lb *1081kg*.
Performance: Maximum speed 102.5mph *165kph* at 16,400ft *5000m*; time to 3280ft *1000m*, 5.3min; service ceiling 24,600ft *7500m*; endurance 2.66hr.
Armament: One fixed Spandau machine gun and one flexibly mounted Parabellum machine gun.
Service: First flight late 1917; service entry early 1918.

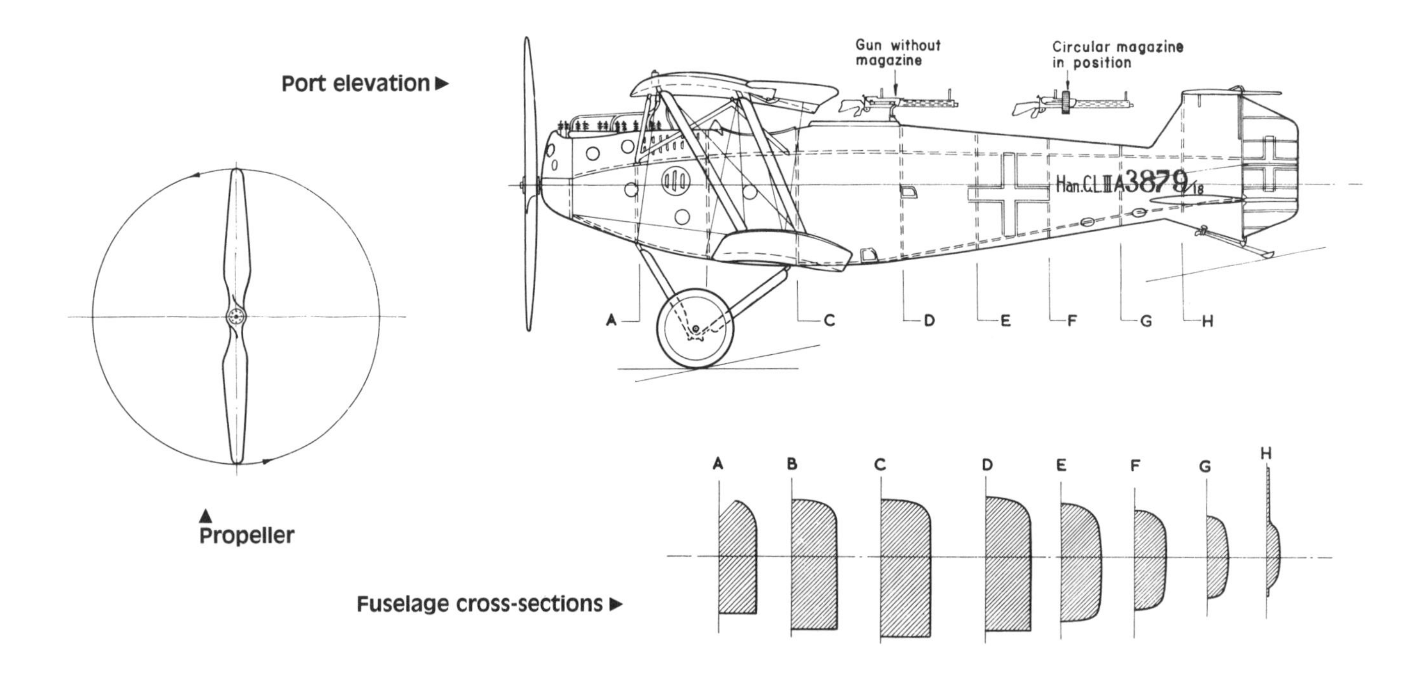

Colour notes
Fabric-covered surfaces were covered with lozenge printed linen, usually of 'B' type pattern in shades of indigo, dark cobalt, sage green, yellow ochre and maroon. The undersurface colours were similar but several shades lighter: a creamy shade was substituted for the ochre, pale violet for indigo, cerulean blue for cobalt, leaf green for sage green and pink for maroon. The rudder was usually covered with plain unbleached linen. Although fuselages were additionally fabric-covered, they were also often daubed with small patches of dark green, mauve and brown colours, which resulted in a generally dark greenish impression. Metal panels and struts were usually dark green or grey. Both patee and Greek crosses – according to the period of service – were painted on Hannovers in the usual locations, with white outlines but not on square white backgrounds. Serial numbers, located as shown, were black. An individual identity device was frequently painted on the fuselage side, often a black numeral on a white band or sometimes fancifully outlined.

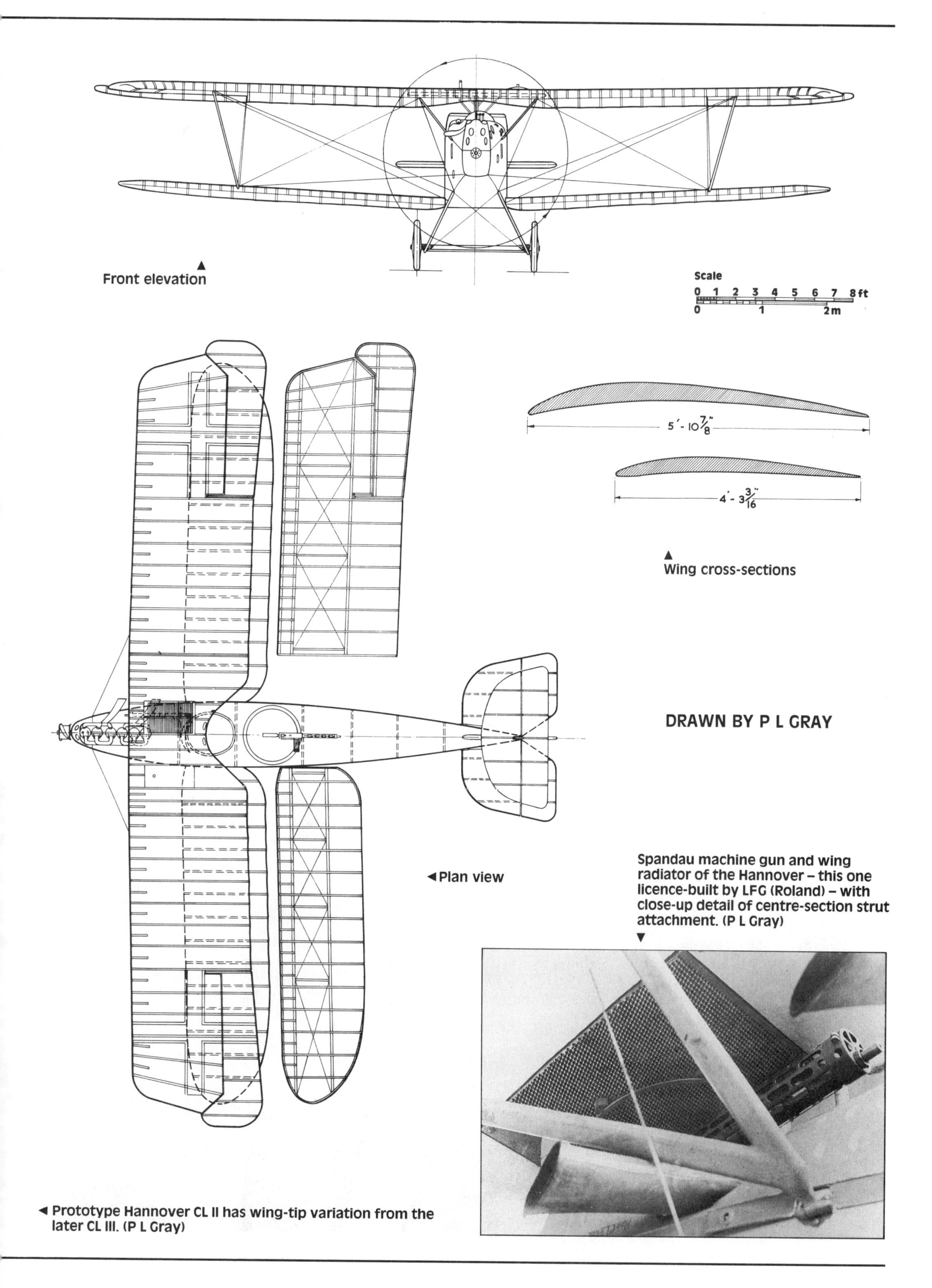

Spandau machine gun and wing radiator of the Hannover – this one licence-built by LFG (Roland) – with close-up detail of centre-section strut attachment. (P L Gray)

Prototype Hannover CL II has wing-tip variation from the later CL III. (P L Gray)

Pfalz D III and IIIa

Country of origin: Germany.
Type: Single-seat scout.
Powerplant: One Mercedes D III six-cylinder, liquid-cooled engine rated at 160hp, (IIIa) D IIIa rated at 180hp.
Dimensions: Wing span 30ft 10in *9.40m*; length 22ft 9½in *6.95m*, (IIIa) 23ft 2in *7.06m*; height 8ft 9in *2.67m*; wing area 238.6 sq ft *22.17m²*.
Weights: Empty 1599lb *725kg*; loaded 1996lb *905kg*.
Performance: Maximum speed 102.5mph *165kph* at 10,000ft *3050m*; time to 3280ft *1000m*, 3.25min; service ceiling 17,000ft *5180m*; endurance about 2.5hr.
Armament: Two fixed Spandau machine guns.
Service: First flight summer 1917, (IIIa) January 1918.

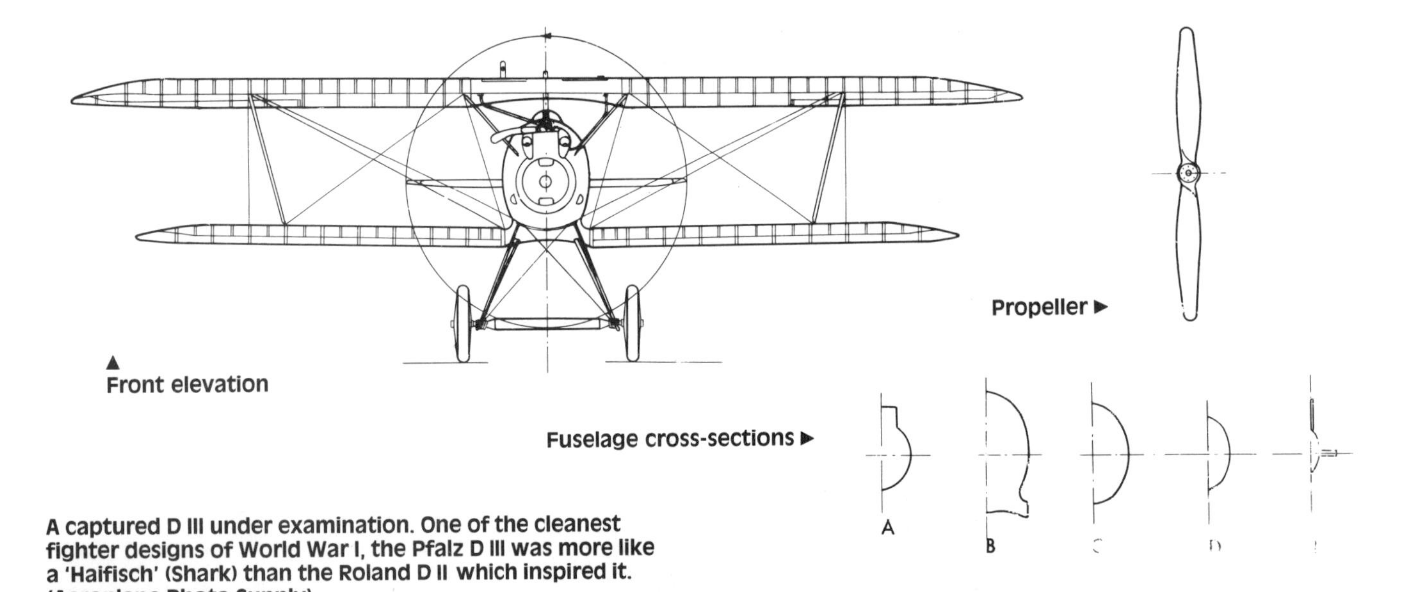

A captured D III under examination. One of the cleanest fighter designs of World War I, the Pfalz D III was more like a 'Haifisch' (Shark) than the Roland D II which inspired it. (Aeroplane Photo Supply)
▼

Colour notes
Practically all D IIIs were sent to the aircraft parks finished in aluminium dope overall. Interplane and centre-section struts were natural varnished spruce, and propellers were usually of walnut and ash laminations, clear varnished. *Jasta* aircraft were decorated with unit and individual insignia, those of *Jasta 10* having the complete tail assembly, the nose (as far aft as the front undercarriage struts), wheel discs and struts painted deep chrome yellow. D III 1370/17 of this unit was flown by *Lt* Klein and had the fuselage cross flanked by light-coloured bands approximately the same width as the cross. Later-production D IIIs had their wings covered in lozenge printed fabric, the remainder of the airframe still being aluminium. Straight-sided Greek crosses enclosed by 15cm white surrounds were introduced on 15 April 1918, while from 16 May a narrow white outline to the sides only of the crosses was ordered.

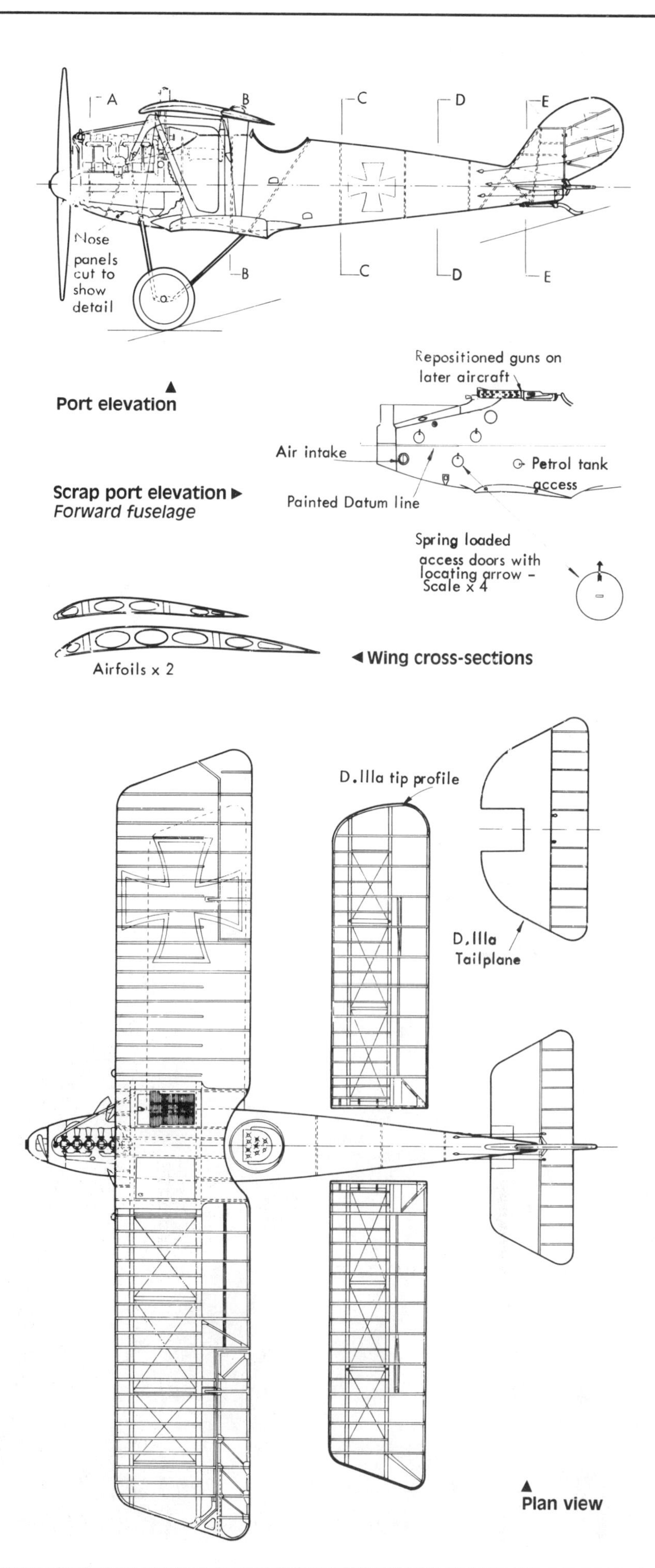

DRAWN BY P L GRAY

Pfalz D XII

Country of origin: Germany.
Type: Single-seat scout.
Powerplant: One Mercedes D IIIa six-cylinder, liquid-cooled engine rated at 160hp.
Dimensions: 29ft 6¼in *9.00m*; length 20ft 10in *6.35m*; height 8ft 10¼in *2.70m*; wing area 233.5 sq ft *21.7m²*.
Weights: Empty 1579lb *716kg*; loaded 1978lb *897kg*.
Performance: Maximum speed 105.5mph *170kph*; time to 3280ft *1000m*, 3.4min; service ceiling 18,500ft *5640m*; endurance 2.5hr.
Armament: Two fixed Spandau machine guns.
Service: Service entry early 1918.

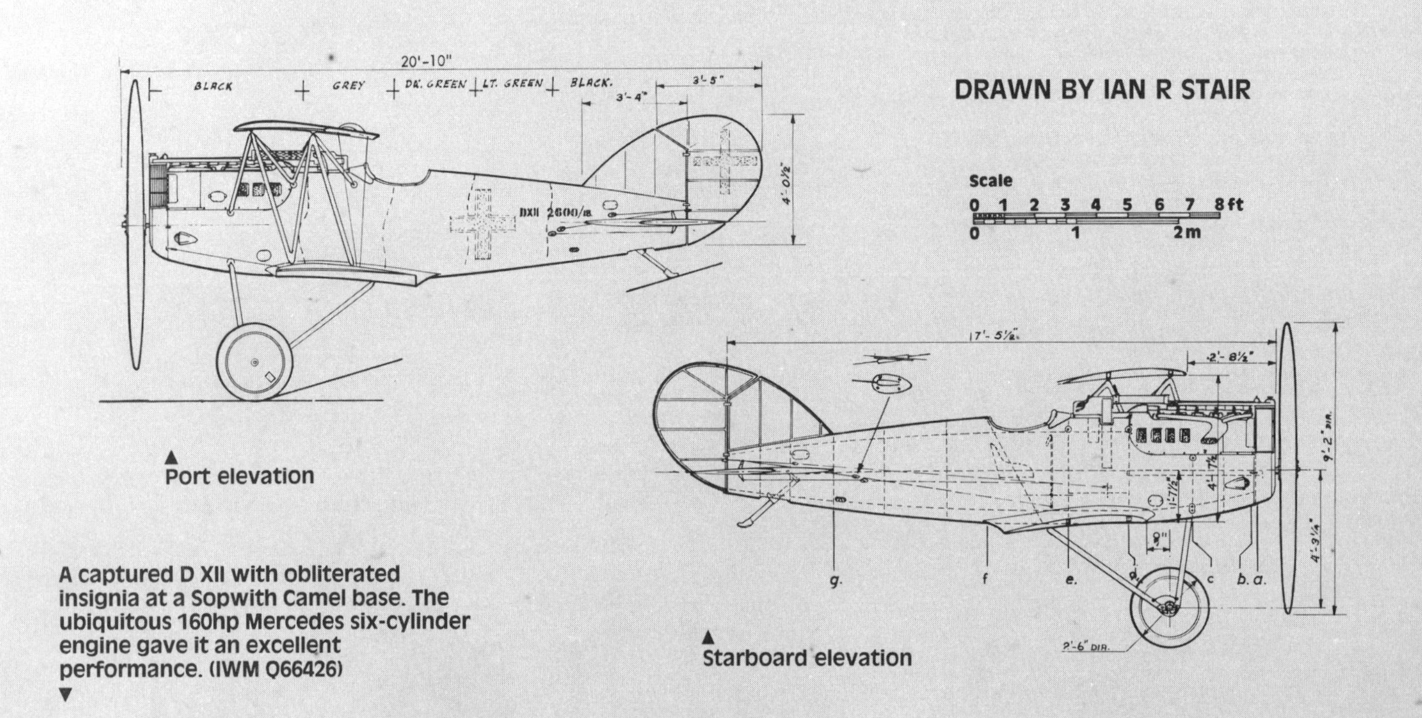

▲ Port elevation

▲ Starboard elevation

A captured D XII with obliterated insignia at a Sopwith Camel base. The ubiquitous 160hp Mercedes six-cylinder engine gave it an excellent performance. (IWM Q66426)
▼

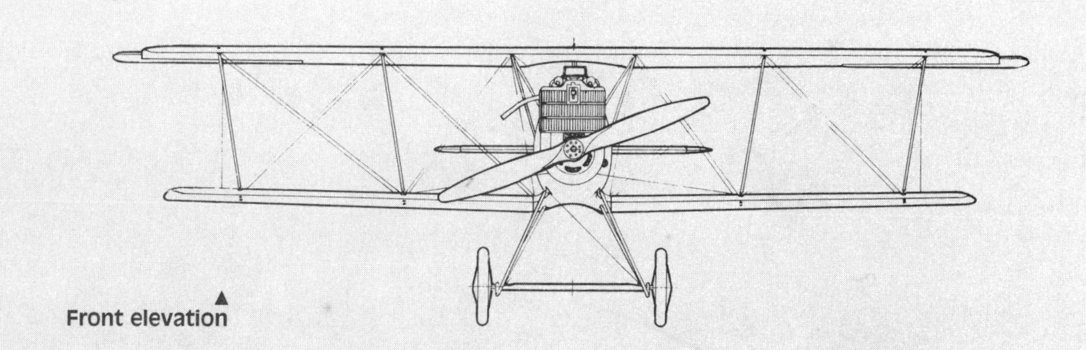

Front elevation

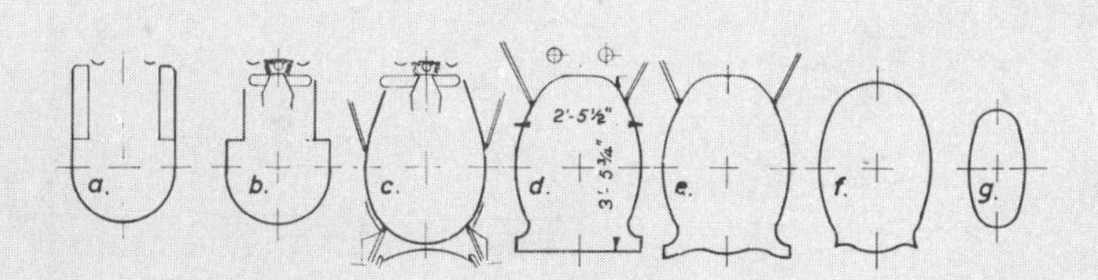

Fuselage cross-sections

Rigging notes
Dihedral: Top wing nil, lower wing 1°.
Incidence: Top wing centre-section 4½°, tip 3°; lower wing root 4°. tip 3°.

▲
Lozenge-pattern fabric on the wings and a varnished ply fuselage became standard for the D XII. Although small, it used double-bay interplane struts and robust bracing. (H J Norwarra)

A Pfalz D XII in original camouflage, with tones faded into bands. It was an exceptionally strong fighter, but few survive in preservation today. (IWM Q66073)
▼

Colour notes

Fuselage, fin and tailplane – undersurfaces varnished plywood, uppersurfaces camouflaged as shown on side view. Wings and elevators – lozenge pattern camouflage as follows:
Uppersurfaces: 1. Dark blue 7-086. **2.** Deep violet 0-014. **3.** Greyish-green 5-064. **4.** Tan 3-044. **5.** Grey-blue 7-085.
Undersurfaces: 1. Light grey-blue 7-080. **2.** Bright pink 8-091. **3.** Dull pink 1-021. **4.** Buff 3-043. **5.** Light green 6-072.
British Standard Colour references quoted as a guide only – they tend to be more definite than the actual colours used.
Rudder – white. Struts and undercarriage – brown.

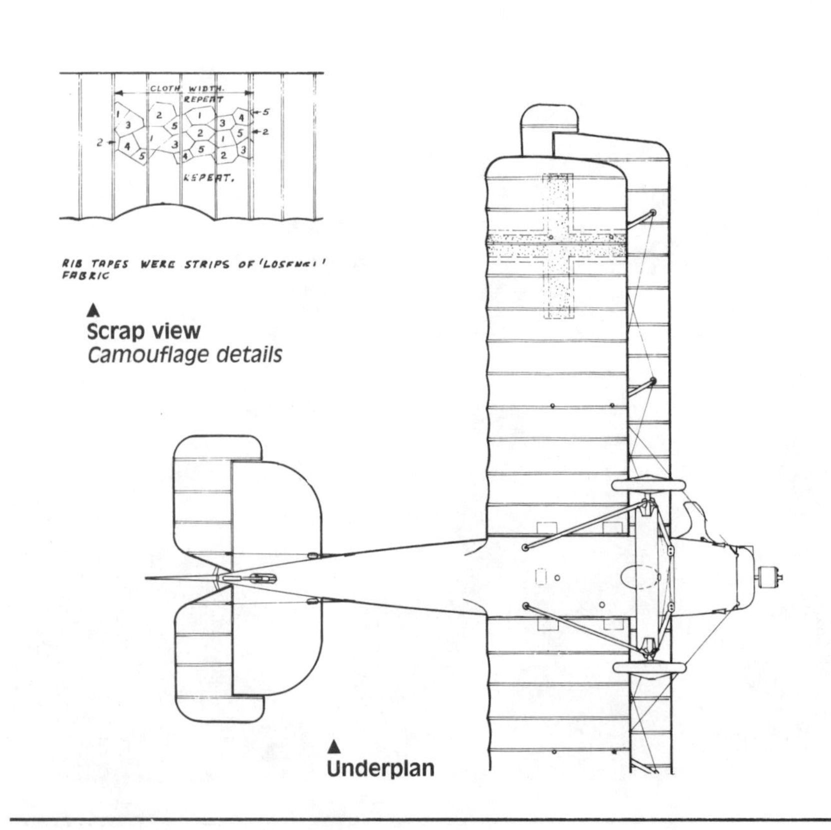

▲
Scrap view
Camouflage details

▲
Underplan

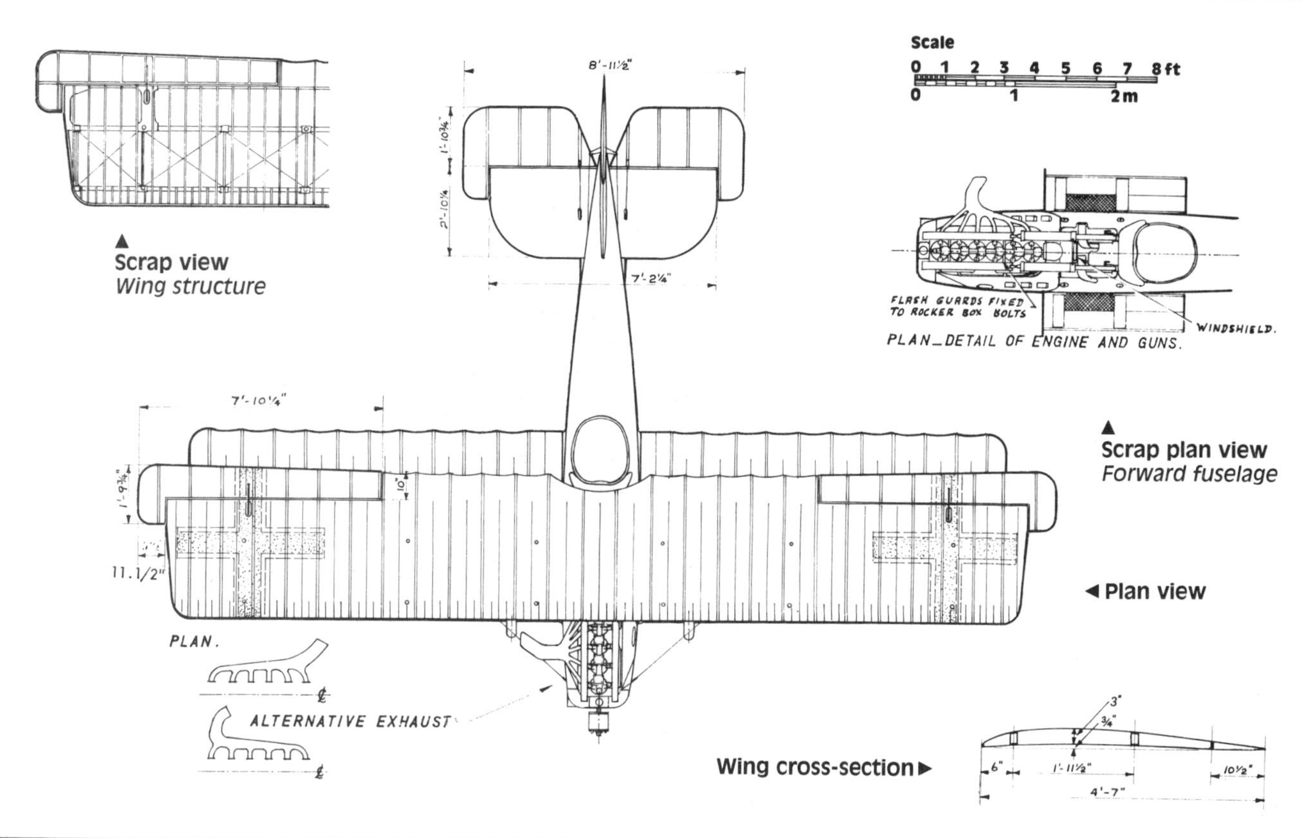

▲ **Scrap view**
Wing structure

▲ **Scrap plan view**
Forward fuselage

◄ **Plan view**

Wing cross-section ►

Roland D VIb

Country of origin: Germany.
Type: Single-seat scout.
Powerplant: One Benz Bz IIIa six-cylinder, liquid-cooled engine rated at 200hp.
Dimensions: Wing span 30ft 10in *9.40m*; length 20ft 9in *6.32m*; height 9ft 2¼in *2.80m*; wing area 238.2 sq ft *22.13m²*.
Weights: Empty 1411lb *640kg*; loaded 1896lb *860kg*.
Performance: Maximum speed 125mph *201kph*; time to 3280ft *1000m*, 2.5min; service ceiling 21,000ft *6400m*; endurance 2hr.
Armament: Two fixed Spandau machine guns.
Service: First flight 1918.

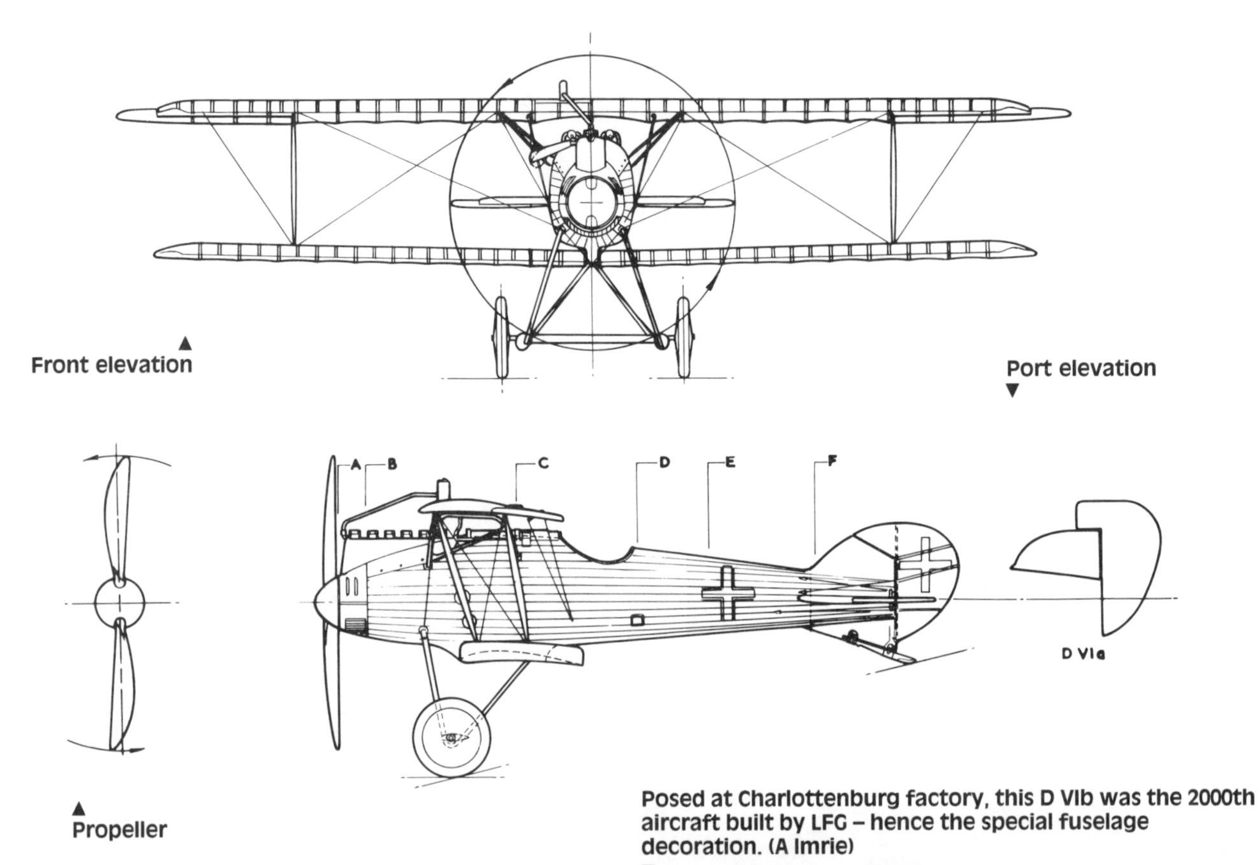

Posed at Charlottenburg factory, this D VIb was the 2000th aircraft built by LFG – hence the special fuselage decoration. (A Imrie)

Colour notes
Ex-works, D VIbs had wings and tailplane covered in lozenge printed fabric dyed in shades of dark green, light green, maroon, indigo and ochre on the top surfaces; the undersurfaces bore similar shades, but of considerably lighter hue, e.g. pink replaced maroon, cobalt replaced indigo. Rib tapes were of a light blue shade, and the rudder was covered with plain linen fabric. The 'clinker' strips of the fuselage were in their natural spruce finish, as was the ply-covered fin; both were well covered with a transparent protective varnish, which resulted in a warm straw shade. The spruce interplane struts were the same colour, but undercarriage struts, centre-section struts and metal nose panels were painted medium grey. Wheel discs, of light gauge metal, were lozenge painted to match the fabric.

▲
Experimental D VIb with double-bay 'I' struts. Note camouflage fabric on wheel covers of an otherwise natural finish machine. (A Imrie)

DRAWN BY P L GRAY

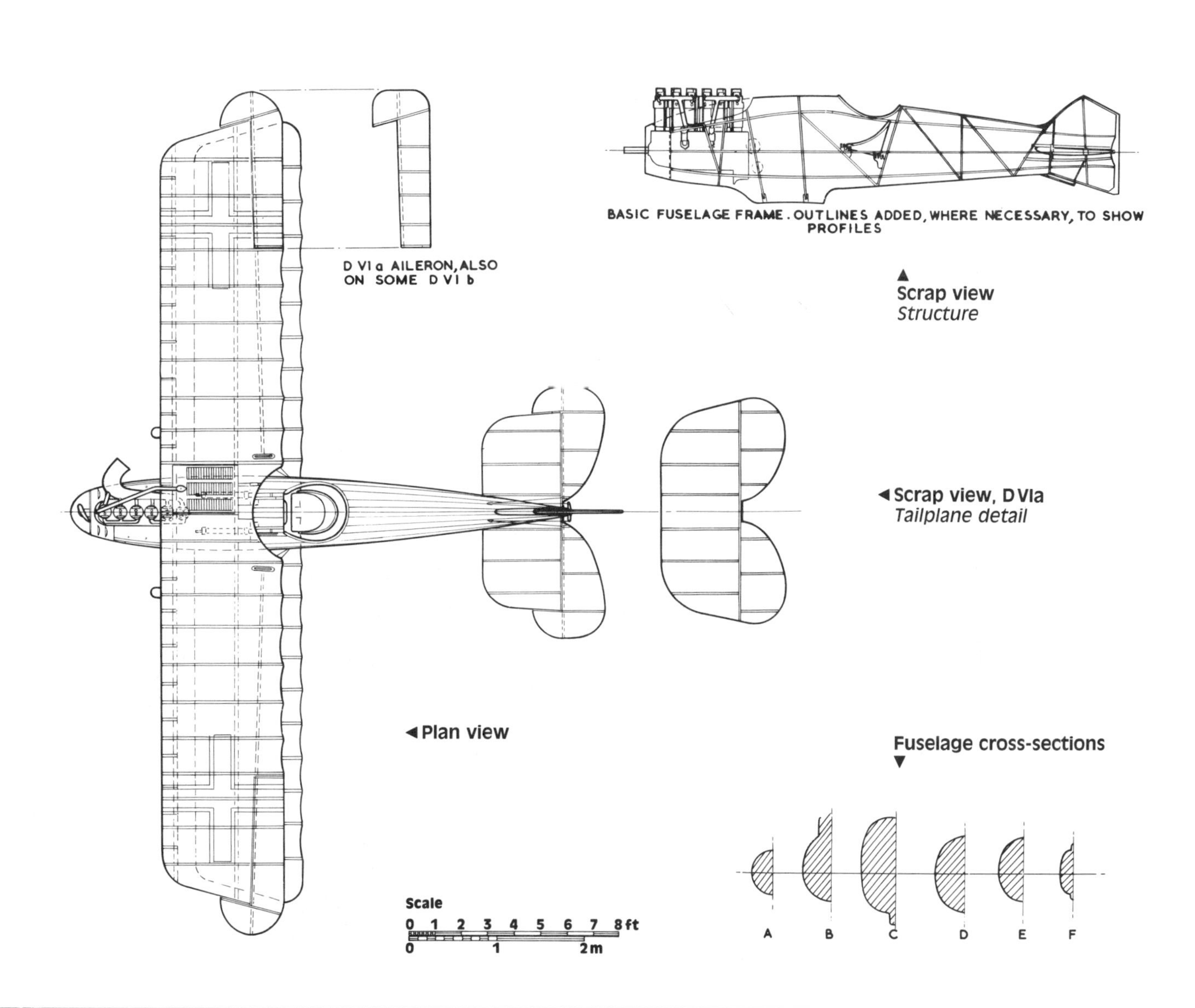

▲
Scrap view
Structure

◄ Scrap view, D VIa
Tailplane detail

◄ Plan view

Fuselage cross-sections
▼

Etrich Taube

Country of origin: Austria-Hungary.
Type: Two-seat reconnaissance aircraft.
Powerplant: One Mercedes six-cylinder, liquid-cooled engine rated at 100–120hp.
Dimensions: Wing span 47ft 0½in *14.34m*; length 32ft 3¾in *9.85m*.
Weights: Empty about 1920lb *870kg*.
Performance: Maximum speed 62mph *100kph*; time to 2600ft *800m*, 15min.
Armament: None.
Service: Service entry 1913.

Note
The drawings depict a Taube in prewar civil configuration.

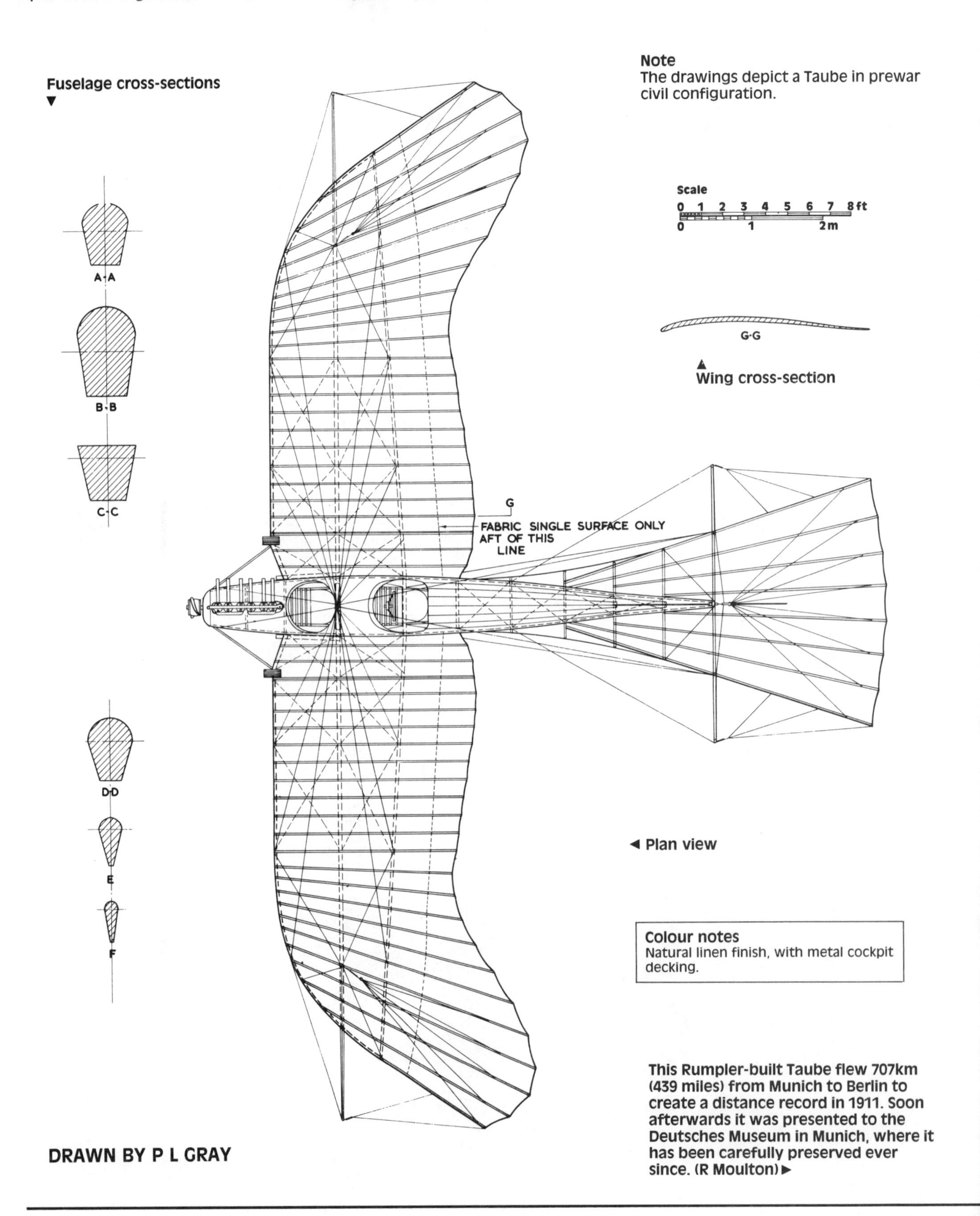

Colour notes
Natural linen finish, with metal cockpit decking.

This Rumpler-built Taube flew 707km (439 miles) from Munich to Berlin to create a distance record in 1911. Soon afterwards it was presented to the Deutsches Museum in Munich, where it has been carefully preserved ever since. (R Moulton) ▶

DRAWN BY P L GRAY

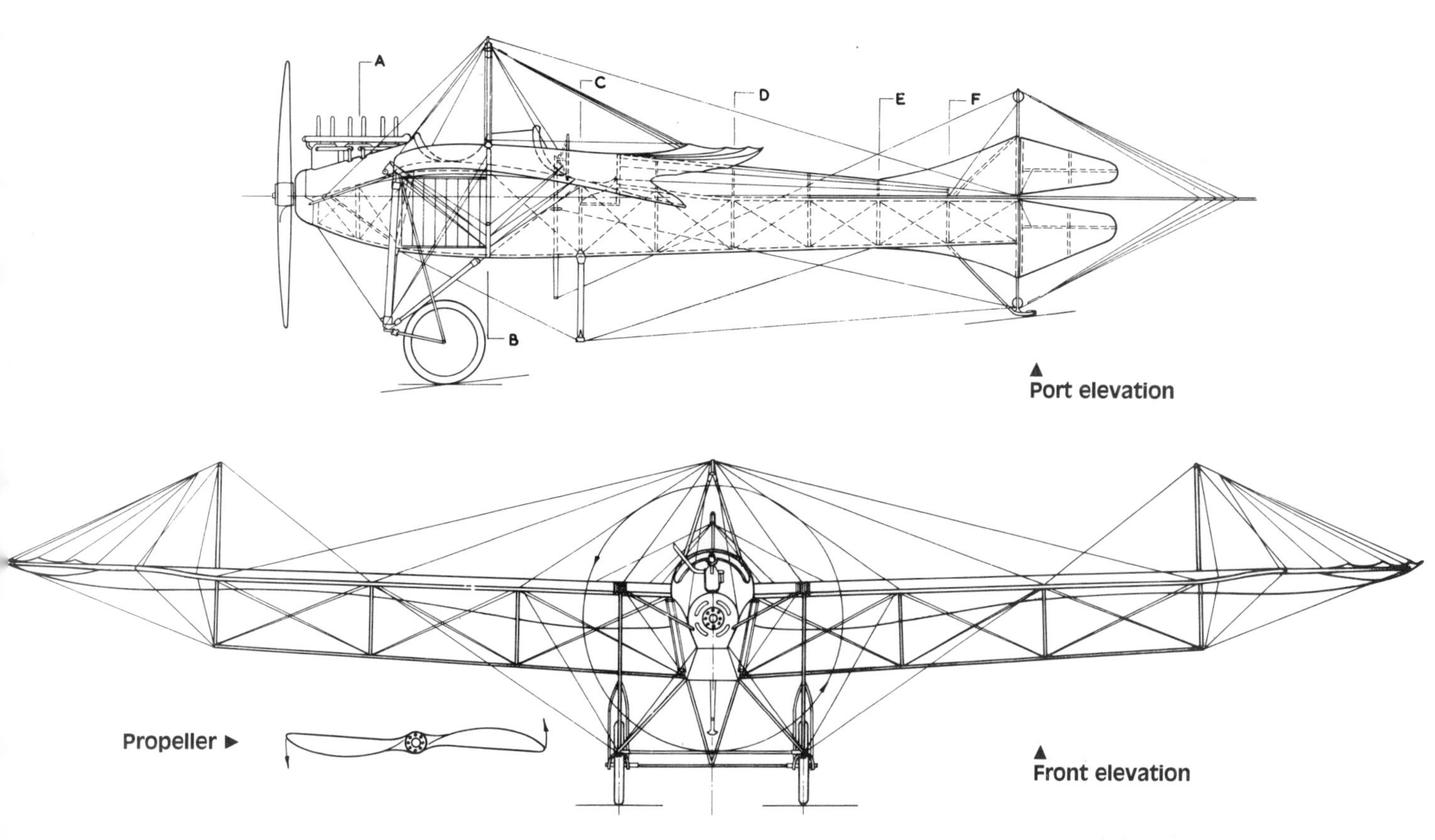
A
B
C
D
E
F
Port elevation
Propeller
Front elevation

Hansa-Brandenburg D I

Country of origin: Austria-Hungary.
Type: Single-seat scout.
Powerplant: One Austro-Daimler liquid-cooled engine rated at 150hp, 160hp or 185hp.
Dimensions: Wing span 27ft 10½in *8.50m*; length 20ft 8in *6.30m*; height 9ft 1¾in *2.79m*.
Weights: Empty 1499lb *680kg*; loaded 2050lb *930kg*.
Performance: Maximum speed 112mph *180kph*, (185hp engine) 116.5mph *187.5kph*; time to 3280ft *1000m*, 4min; endurance about 2hr.
Armament: One fixed Schwarzlose machine gun.
Service: First flight summer 1916.

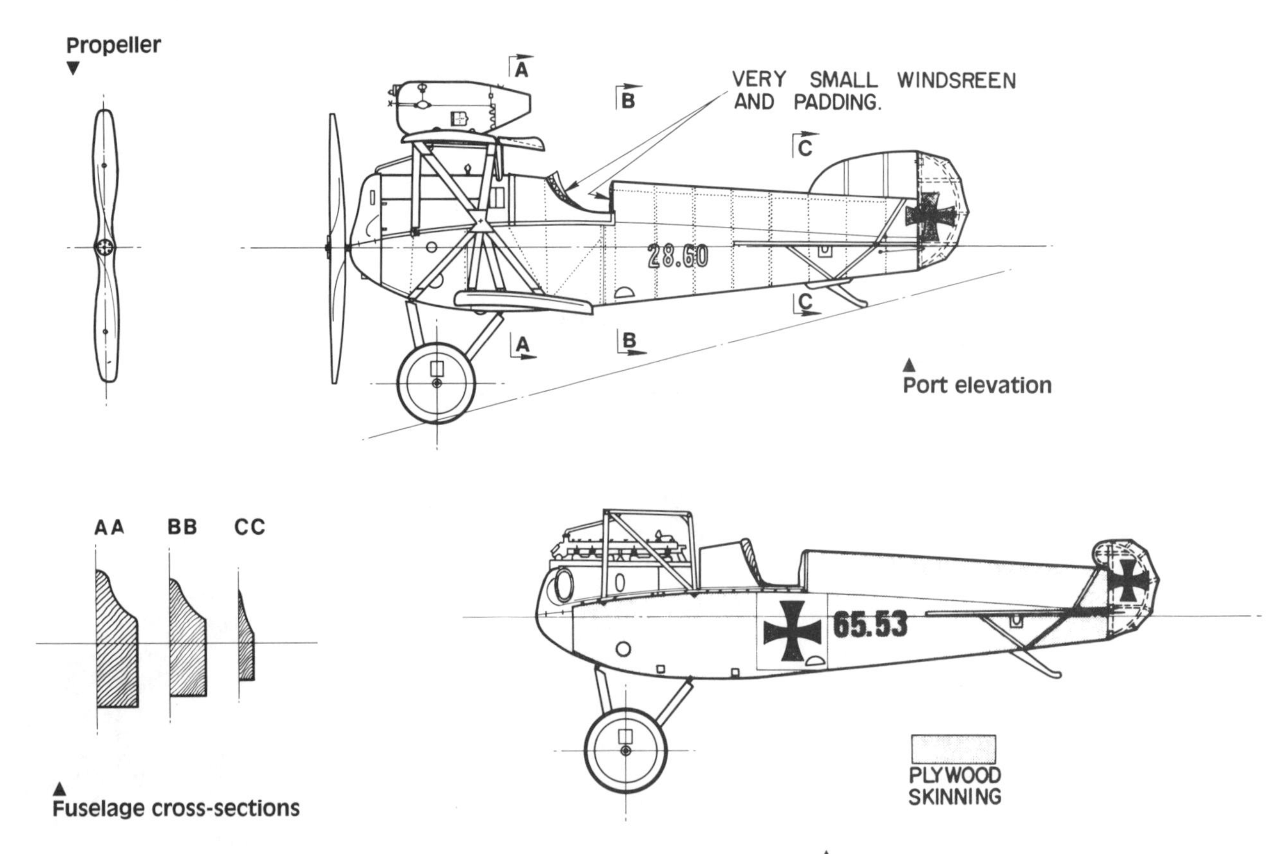

▲ Port elevation

▲ Fuselage cross-sections

▲ **Port elevation**
Flown by Hauptmann *Godwin Brumowski*, Flik 41, *1917*

Two Brandenburg D Is of *Flik 4* on the Isonzo Front in February 1917. KD serial 65:50 (right) has a 160hp and 65:74 (left) a 150hp Austro-Daimler motor. (B Toschinger/ H Woodman)
▼

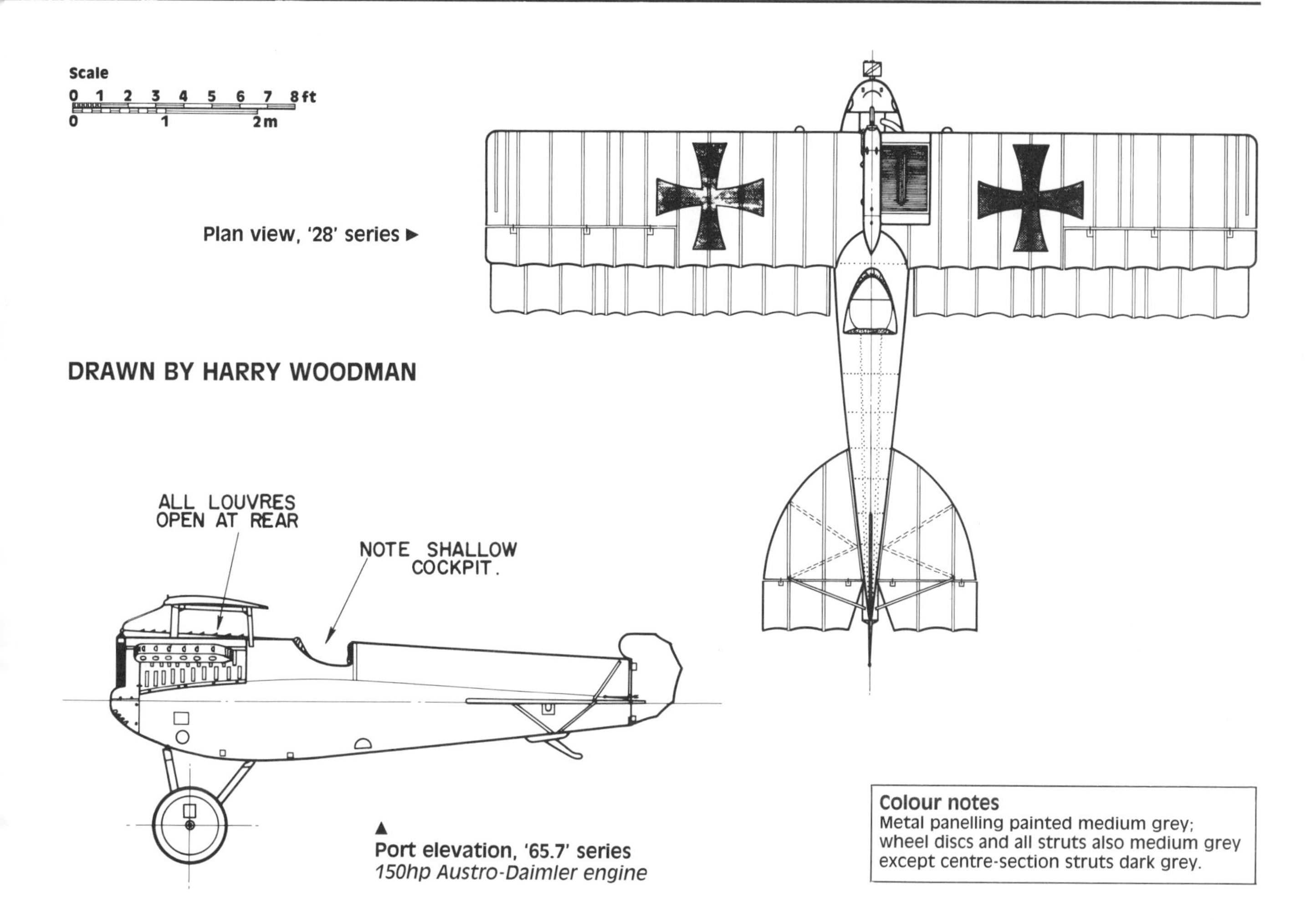

A Phönix-built D1, serial 28:28, showing to best advantage the interplane strut arrangement which eliminated wire bracing. (G Haddow/H Woodman)
▼

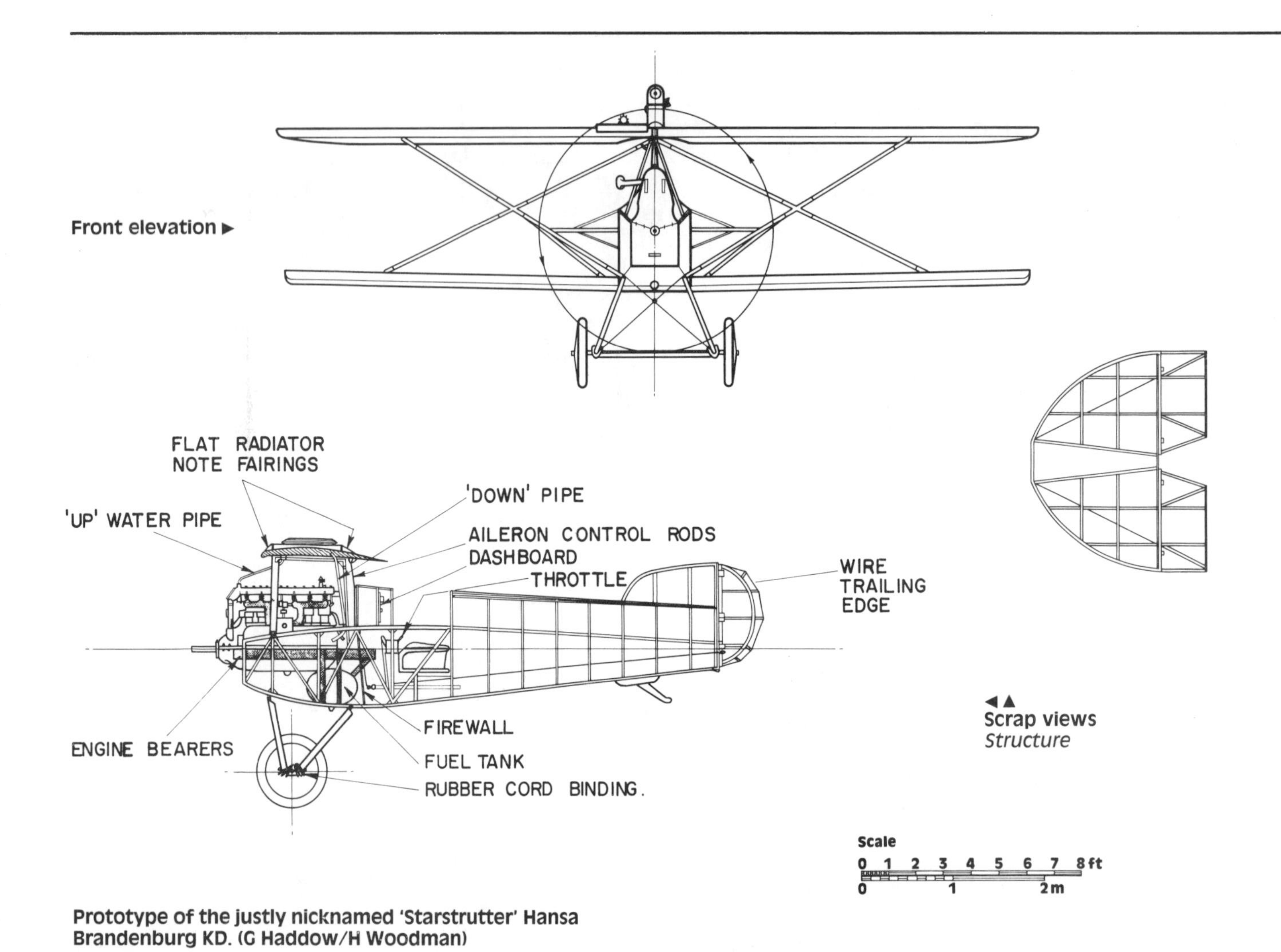

◄▲
Scrap views
Structure

Prototype of the justly nicknamed 'Starstrutter' Hansa Brandenburg KD. (G Haddow/H Woodman)
▼

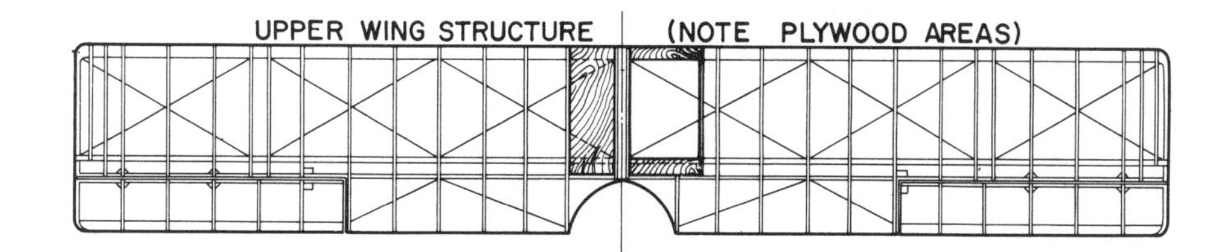

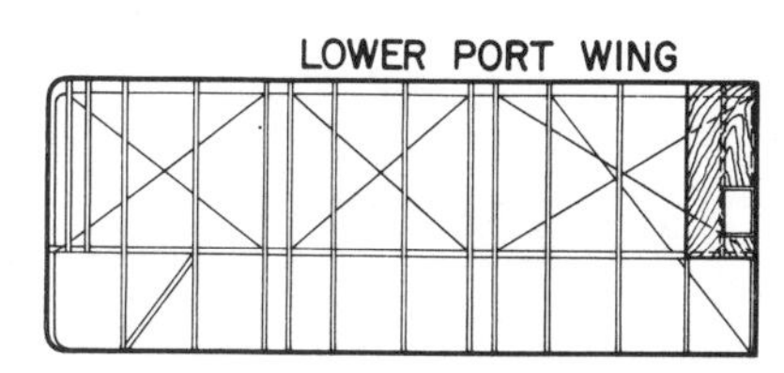

◄ **Scrap plan views**
Wing structure

Instrument panel of a Phönix-built D I. The layout was considerably in advance of many contemporaries. (H Woodman)
▼

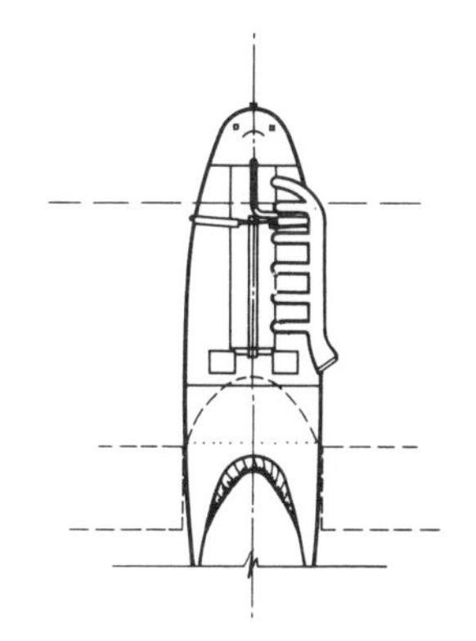

◄ **Scrap plan view**
Forward fuselage

Scrap plan view ►
Lower wing

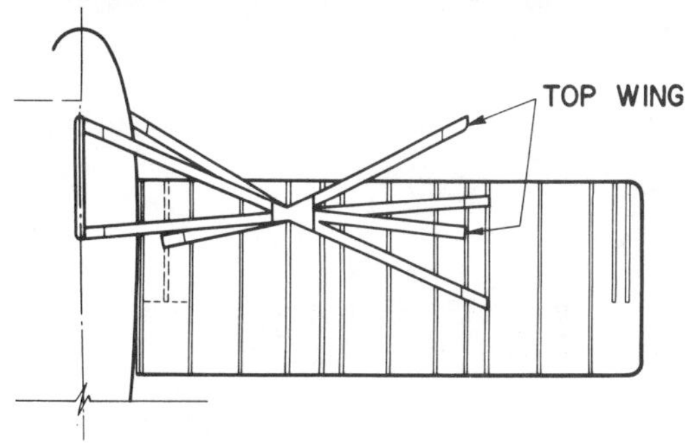

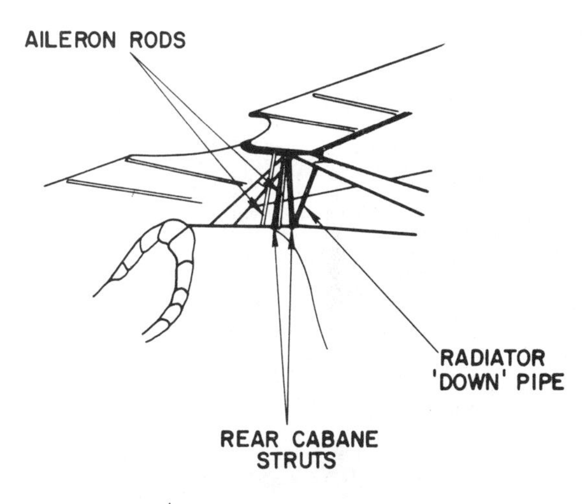

▲
Scrap view
Layout of centre-section

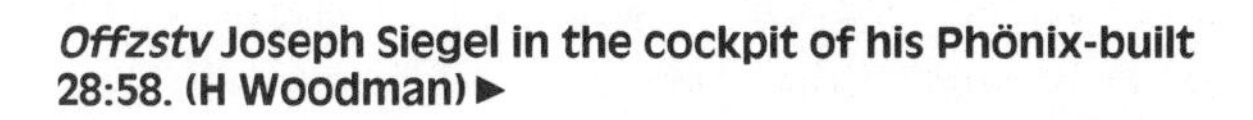

Offzstv **Joseph Siegel in the cockpit of his Phönix-built 28:58. (H Woodman)** ►

The Publisher wishes to thank the following draughtsmen whose drawings appear in this volume

GEORGE COX
P L GRAY
DAVID R JONES
BJÖRN KARLSTRÖM
KENNETH McDONOUGH
F PAWLOWICZ
IAN R STAIR
HARRY WOODMAN